SpringerBriefs in Intelligent Systems

Artificial Intelligence, Multiagent Systems, and Cognitive Robotics

Series editors

Gerhard Weiss, Maastricht, The Netherlands
Karl Tuyls, Liverpool, UK

AF168296

More information about this series at http://www.springer.com/series/11845

Malachy Eaton

Evolutionary Humanoid Robotics

 Springer

Malachy Eaton
Department of Computer Science
 and Information Systems
University of Limerick
Limerick
Ireland

ISSN 2196-548X   ISSN 2196-5498   (electronic)
SpringerBriefs in Intelligent Systems
ISBN 978-3-662-44598-3   ISBN 978-3-662-44599-0   (eBook)
DOI 10.1007/978-3-662-44599-0

Library of Congress Control Number: 2014959413

Springer Heidelberg New York Dordrecht London

Printed on acid-free paper

Springer-Verlag GmbH Berlin Heidelberg is part of Springer Science+Business Media
(www.springer.com)

Preface and Acknowledgments

In writing this book I must first pay tribute to two leading researchers in the evolutionary robotics field, Dario Floreano and Stefano Nolfi, whose book *Evolutionary Robotics* formed the de facto reference and touchstone for researchers in the evolutionary robotics domain since its publication in 2000. In particular I would like to thank Prof. Dario Floreano, the director of the Laboratory of Intelligent Systems at the École Polytechnique Fédérale de Lausanne (EPFL) in Switzerland. I had the very good fortune to serve a short sabbatical as a guest researcher with Dario and his group in the "Pavilion Jaune" at EPFL in the winter of 2000/2001. Indeed, it was at precisely this time that Dario and Stefano's book was to be published, and I remember at least every second day going across to the bookshop at EPFL to see if a copy was available. Alas, due to unforeseen publishing delays I was not able to purchase my own copy until not long before the end of my all-too-short sabbatical stay. Dario, with great grace, wrote a short dedication, together with Stefano:

> We hope that you will find this book stimulating and develop this approach much further.

While I cannot claim to have made major advances, or to have taken this approach that much further, things have moved on quite a bit since 2001. I sincerely hope that in this book I have taken this advice to heart, in describing my own and other researchers' work in the application of evolutionary techniques to the ever more important and prevalent domain of humanoid robotics. I follow on, to a certain extent, where their text left off in the penultimate chapter of their book entitled "Complex Hardware Morphologies: Walking Machines". The focus of this chapter was, however, on 4-, 6-, and 8-legged locomotion rather than bipedal locomotion that most "humanoid" robots use.

I would also like to thank Prof. Dave Cliff, now at the University of Bristol, but who was at the University of Sussex when I met him first at a mechatronics conference in Halmstad, Sweden in 1993. At this stage I had just completed my Ph.D. thesis entitled "Genetic Algorithms and Neural Networks for Control Applications", and was actively interested in looking at ways in which the neural and, in particular, the evolutionary approach could be applied in the field of mobile robots. My thesis had been concerned mainly with the application of these techniques to theoretically defined problems in the control area, where systems were precisely specified by differential equations of different orders, sometimes incorporating time delays of varying magnitudes. The results obtained using the neuro-genetic approach were then compared to those obtained using either theoretically derived time-optimal solutions, or those using empirically derived control parameters such as the Ziegler–Nichols and refined Ziegler–Nichols rules for Proportional-Integral-Derivative (PID) control systems.

Dave Cliff's work (along with that of his colleagues at the University of Sussex) opened up a whole new domain for the application of this new neuro-genetic approach to control applications which were not so precisely specified, and which could incorporate many degrees of freedom, and potentially operate successfully in real-world, noisy environments. I had the good fortune to have Dave accept an invitation to the University of Limerick in the winter of 1993, where he gave a talk entitled "Evolving Visually Guided Robots". This talk was very well received and his research and that of his colleagues at the University of Sussex formed one of the main motivations for my subsequent research interests in the field of evolutionary robotics, as it had recently become known.

Moving closer to the present time, I would like to express my sincere thanks to Ronan Nugent of Springer, without whose enthusiasm this project would never have got off the ground, and to also extend my appreciation to the Series Editors for their helpful and constructive comments. Also, I would like to thank my head of department, Annette McElligott, for her support throughout the enterprise, and finally my wife, Patricia, for her understanding through what must, at times, have seemed like a neverending process.

There is little that is actually new in this text. What I have tried to do is to collect the various strands encompassing the fields in this diverse and rapidly evolving subject, and to hopefully present them in a reasonably coherent and concise manner to the moderately educated reader. Where mathematical formulae or intricacies are perceived as essential to the presentation of the topic these are included, but efforts have been made to avoid any unnecessary mathematical complications in order to make the text accessible to as wide an audience as possible.

Although I have endeavoured to be as objective as possible in my treatment of the subject matter herein, as in all works of this nature, certain preferences and biases must, of their nature, creep in. Of course, a text of this size cannot purport to be comprehensive. However, a representative cross-section of references to the most current material, together with material of historical interest, is given, and pointers to the literature are provided at frequent intervals in the text.

Intended Audience

This book is intended to be of use to the following categories of readers:

– Researchers looking for an up-to-date and concise review of key aspects in the state of the art in the field of evolutionary robotics in general, following on from Stefano Nolfi and Dario Floreano's groundbreaking text *Evolutionary Robotics*
– Researchers involved in the evolutionary robotics field who require a brief introduction to the humanoid robotics area and how they might apply their expertise to this domain
– Researchers involved in the humanoid robotics field who are curious about how evolutionary robotics might have some applications in their area
– Researchers in the biological sciences field interested in recent advances in this bioinspired area of research
– Researchers who are already involved in the EHR area who would like a concise reference "handbook" to their field with a comprehensive set of references together with a concise summary of major strands of research in their field from its inception to the present time
– Educators at the advanced undergraduate/postgraduate level who require an up-to-date concise introduction to/survey of the field of evolutionary robotics or humanoid robotics
– Educators at the postgraduate/advanced postgraduate level who want a text specifically in the area of the application of evolutionary techniques to humanoid robots, tying together all of the most recent research in this field in a cohesive manner, and by a single author
– The general informed (and reasonably well educated) reader, who realises that certain issues raised in this book's content may have significant implications on society within their lifetimes and beyond.

Limerick, September 2014   Malachy Eaton

Contents

Chapter 1
Introduction

1.1 Scope of the Book

In this text we look to the past to two distinct strands of research into autonomous robots, evolutionary robotics and humanoid robot research, and how these strands are now beginning to converge in the novel field of evolutionary humanoid robotics. We investigate some of the current and emerging work in this new and exciting field. We address briefly some of the motivations. Why evolve robot bodies or brains, rather than go through a rigorous design process? And why should we have a particular interest in the creation of specifically humanoid robots, rather than, say, wheeled robots, or four-legged (quadrupedal) designs?

Following a brief overview of some recent developments in intelligent autonomous robotics, we discuss the field of evolutionary algorithms, looking to recent developments in the evolutionary robotics field. We discuss some of the issues involved in the evolution of mobile robots both in simulation and on real robots and the associated "reality gap" issue. Following a discussion of early research in the field, from the early 1990s, we move on to look at the state of the art in evolutionary humanoid robotics. We then address the important, but often overlooked area of the performance evaluation and benchmarking of autonomous robots, and humanoid robots in particular.

Finally we look briefly to the future—a future where humankind may be dominated, or even exterminated, by a vastly superior species of evolved humanoid, or post humanoid robots. Or, as Minsky (1970) is once famously quoted as saying in an interview with Life magazine, "If we are lucky they might decide to keep us as pets".

On the other hand, we might envisage a future where mankind and highly evolved humanoid robots work side by side in a semiutopian fashion, where

© The Author(s) 2015
M. Eaton, *Evolutionary Humanoid Robotics*,
SpringerBriefs in Intelligent Systems, DOI 10.1007/978-3-662-44599-0_1

poverty is a thing of the past, all back-breaking and menial tasks will be undertaken by our robot companions, and humankind will be left free to enjoy lives of unbridled bliss.

1.2 Possible Approaches to This Text

I toyed with several approaches that were helpfully suggested for the writing of this book. One approach involved the production of an edited book, created by soliciting contributions from different authors in the twin areas of evolutionary robotics and humanoid robotics, and the subsequent collation of these contributions into a single edited volume. Although this approach had its attractions, I felt that it might not result in a cohesive text that would specifically address the core topic of evolutionary humanoid robotics.

There are books that cover the diverse fields of evolutionary algorithms, robot learning, humanoid robotics, and even specifically the topic of evolutionary robotics. However, there is no one text that addresses the specific application of evolutionary techniques to the design of humanoid robots, which is the particular focus of this text. We will attempt to synthesise all of these ideas, insofar as they relate to the specific and novel field of evolutionary humanoid robotics (EHR). This book also provides a summary of the most up-to-date results and developments in this rapidly evolving area.

There exist other volumes of collated contributions, notably the monumental Springer *Handbook of Robotics*, which, although it does not address the topic of evolutionary humanoid robotics explicitly, is a very useful background reference, and includes further details relating to the areas of "classical" robotics for researchers in this area (Siciliano and Khatib 2008). Soon to be published in a new edition, this weighty tome (over 1,500 pages) contains a wealth of information on all things robotics, including no fewer than four independent forewords, one of which was written by the "father of nouvelle AI", Rodney Brooks. This text also contains excellent introductory chapters on the separate fields of humanoid robotics and evolutionary robotics. I recommend that engrossing work as reading matter in conjunction with this book, together with Springer's equally weighty *Handbook of Automation*, which contains a chapter discussing the use of evolutionary techniques for automation (Nof 2009).

So it was decided, for the reasons given, to go with an authored text, and what you have before you is the result of this work. Although this is a relatively slender tome, it represents the distillation of several hundred research articles, and quite a number of texts.

Some of the research described relates to my own modest research efforts in the field, for which I make no apologies. I am more familiar with my own work than with that of other researchers so this makes sense from an author's perspective. This does not, in any sense, imply that this research is of more relevance or overall importance to the general field.

A luxury afforded to me in my early experiments on the evolution of controllers for theoretically defined systems was, in many cases, the availability of some reference point, for example, a time-optimal controller, with which to compare my genetically derived results. Alas, no such luxury is afforded to researchers working with robots with many degrees of freedom and complex system dynamics operating in highly complex and nonstationary real-world environments.

So I will briefly refer in this text to my own relevant research, from initial research in the early 1990s on the evolution of control algorithms for theoretically defined controllers to some of my later results in the evolution of bipedal loco-motion in a simulated (and later embodied) robot with many degrees of freedom. Indeed, Eaton and Davitt (2006, 2007) conducted some early experiments in the application of evolutionary techniques to the control of a humanoid robot with many degrees of freedom (elbows, knees, hips, etc.) and simulated using a simulator employing an accurate physics engine.

I have tried to at least mention as many areas of relevance to this topic as possible with appropriate references to more comprehensive treatments; however, with a book describing the state of the art in a novel and fast-moving field there may well be omissions. For these I apologise. However, I hope that, for the most part, this will prove an informative and perhaps even enjoyable introduction to the field of evolutionary humanoid robotics. So it is hoped that you, gentle reader, will find this, if not exactly an easy read, then one that presents a cohesive synthesis of the work to date in the diverse fields of evolutionary algorithms as applied to robotics, and humanoid robotics in particular, together with some not-too-unreasonable visions of the future possibilities for this fascinating and far-reaching technology.

It is hoped that this book also forms an aid to more advanced researchers in the field by providing a concise synthesis of work to date, and that it also points out areas of specific difficulty or of possible duplication of effort, and in doing so may help to generate a roadmap for future interesting and innovative work in this area.

Space precludes the inclusion of detailed fitness graphs, and of photographs of large numbers of different experimental platforms, etc. This level of detail, in any case, is probably not appropriate in a text of this size. However, in-depth experimental information is generally provided in the original articles, for which I provide detailed references. In addition there are several texts currently available providing excellent photographs and diagrams of the different humanoid (and non-humanoid) robotics platforms in existence today, among which Springer's aforementioned formidable *Handbook of Robotics* stands out. Another, fairly recent, publication dealing with some of the issues we raise in Chap. 8, regarding ethical and societal implications in the development/evolution of autonomous humanoid robots whose actions we may not fully comprehend, is aptly entitled *The Coming Robot Revolution: Expectations and Fears About Intelligent, Humanlike Machines* (Bar-Cohen and Hanson 2009). There are also a variety of survey articles available on some of the different topics discussed in this text; some details of these are given in the relevant sections.

1.3 Outline of the Book

The core research area addressed by this book is at the intersection of the evolutionary computation approach and the field of humanoid robotics (Fig. 1.1). We will explore how evolutionary algorithms can be employed in the design of both the "bodies" and the "brains" of autonomous robots, and of humanoid robots in particular. The EHR field encompasses a range of disciplines, from research into human physiology and behaviour to the process of evolution and the control of systems. Later in this chapter we present a brief outline of some of the core aspects and disciplines associated with the EHR field.

In Chap. 2 we discuss further the main topics addressed in this book, introducing the fields of intelligent robotics and evolutionary algorithms. We look at some of the evolutionary algorithms in current use, focusing especially on those with particular application in the evolutionary robotics field. We finish this chapter with a simple example of evolutionary algorithms applied to the control of a simulated artificial creature moving in a two-dimensional environment.

In Chap. 3 we address the field of evolutionary robotics focusing on some of the key topics associated with this field, including fitness function design, and the scalability of results obtained. We briefly outline some of the perceived failings of the ER field and how these issues are being addressed.

In Chap. 4 we move on to look at research into humanoid robots in particular, and their simulators. We also address some of the approaches taken to ameliorate the "reality gap" issue: that is, what do you do when the results obtained in simulation do not exactly match those obtained on the real robot?

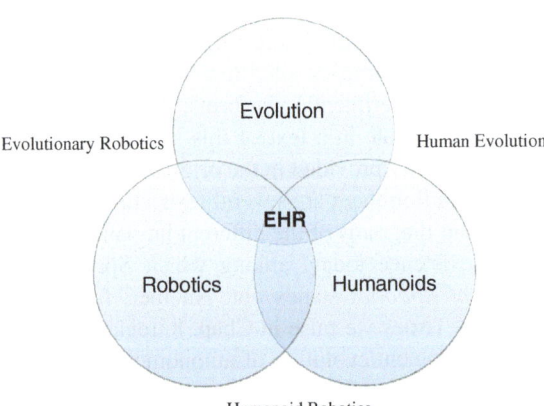

Fig. 1.1 Evolutionary humanoid robotics (*EHR*) lies at the confluence of research into the fields of evolutionary theory and applications, human physiology and behaviour, and autonomous robotics. It also has close ties with the associated fields of evolutionary robotics, human evolution, and humanoid robotics

Chapter 5 addresses the core topic of the book, looking at the different approaches taken to EHR. We introduce some of the main application domains, including bipedal locomotion, dance, and the investigation of human locomotor skills. We then look at initial research in the field—EHR "prehistory", looking at developments from the early 1990s to 2000. The chapter finishes with a simple illustrative example of the evolution of dance.

In Chap. 6 we look at the state of the art in the EHR field, describing developments from 2000 to the present day. This chapter is divided into two sections, initially looking at developments from 2000–2007 followed by an analysis of research from 2008 to the present day. For convenience, the presentation of research in the second section is also given in a tabular format, listing details of the humanoid platform (for embodied experiments), simulator used, application domain, etc. Both sections are further subdivided into simulated and embodied experimentation parts; that is, those experiments conducted entirely in simulation, and those evolutionary experiments in which either some or all of the evolutionary process is conducted on the humanoid robot, or where the evolutionary process takes place entirely in simulation, and is subsequently transferred to a real humanoid robot.

As we move beyond the realm of "proof-of-concept" research on to the area of producing robust and useful humanoid robots aided by the use of evolutionary techniques, the idea of the creation of an evaluation framework, together with benchmarking facilities for robots occupying similar niches in this overall framework, is likely to become an increasingly important issue for future designers. A core issue here is having some method of comparing experiments and results, each of which may employ wildly different hardware platforms, and a wide variety of different evolutionary techniques and associated fitness functions, together with a commensurately wide variation in the amount of domain-specific information supplied. We explore these issues further in Chap. 7, paying particular attention to the current DARPA robotics challenge, and to the ongoing RoboCup initiative.

Then in Chap. 8 we discuss briefly some of the ethical and societal considerations involved in the creation of autonomous agents of broadly humanoid form whose behaviour we do not fully understand, and certain issues that may arise; and we conclude in Chap. 9 with a brief look to the future.

1.4 Brief Discussion of Core Terms: Humanoid, Evolutionary, Robotics

We present here a brief outline of some of the core aspects and disciplines associated with the terms *humanoid*, *evolutionary*, and *robotics*.

Humanoid—in the shape of and/or acting in a fashion similar to a human being (man, woman, or child).
Relevant disciplines: psychology, physiology, anatomy, brain sciences, ethics, etc.

Evolutionary—pertaining to the evolutionary process, as applied to either natural or artificial organisms.
Relevant disciplines: neo-Darwinian evolution, genetics, evolutionary algorithms, etc.
Robotics—pertaining to the design and construction of mechanical devices operating either on command or as a result of being preprogrammed in advance, generally designed to perform some labour-saving function or to perform an entertainment-related role.
Relevant disciplines: control, dynamics, mechanics, etc.

If we look to the origins of the term *evolutionary robotics*, according to Harvey et al. (2005) the term was first used by Cariani in an unpublished paper entitled "Why artificial life needs evolutionary robotics", and later in Cariani's doctoral thesis (Cariani 1989). While research on the application of evolutionary techniques to humanoid robotics can, as we will see, be traced back to the early 1990s, an early (the earliest?) use of the term *evolutionary humanoid robotics* was by Eaton (2007a). I later expanded on this with a paper at AI50—a publication centred around papers submitted to a conference in Monte Verità, Switzerland to commemorate the 50th anniversary of the foundation of Artificial Intelligence (AI) as a distinct academic discipline (Eaton 2007b).

1.5 Very Briefly: Why Evolve Bioinspired/Humanoid Robots?

Why we wish to evolve

(a) robots?
(b) bioinspired robots?
(c) humanoid robots?

- To come up with useful engineering artifacts that might be difficult/impossible to create by other means (a, b, c).
- In the case of (c), to create artifacts that can operate in similar environments to humans, and/or which may also gain more ready acceptance by humans.
- To better understand animal behaviour (b), and/or human behaviour (c).
- To develop useful engineering artifacts to augment human abilities (c), e.g., in the evolution of prosthetic limbs.

1.6 Conclusion

This will not be an uncritical view of the domain of EHR. It is intended as both an introduction to the field, for those unfamiliar with the application of evolutionary algorithms to the robotics area, and to humanoid robots in particular, together with a survey of the state-of-the-art in EHR. As such, we will attempt to present a warts-and-all analysis of what are perceived to be the current problem areas and how (or, indeed, if) they may be tackled in the future.

The ultimate thesis of this text is that the evolutionary approach has many advantages and unique properties as applied to humanoid robot design, certainly in the medium-to-long-term future. However, in order to encompass the full range of human motor functionality and/or cognitive ability (if indeed this is the desired result), then a hybrid approach encompassing elements from evolutionary algorithms, other bioinspired AI techniques such as artificial neural networks and artificial immune systems, and, yes, even Good Old-Fashioned AI (GOFAI) may be desirable, if not essential.

I include the caveat "if indeed this is the desired result" in the above sentence, as I think we (humanity as a whole) are now at the stage where we may need to consider (perhaps on some sort of collective basis) whether the application of ever-advancing strides in technology is actually desirable in every field. This may particularly be the case in areas that have the potential for a profound impact on the lives of humans in the future, such as advanced autonomous robotics including humanoid robotics, nanotechnology, and genetic engineering.

In some cases it appears that technology is driving ever-more advanced technology without necessarily producing more beneficial results for humans, the end-users. It could be argued that mankind is now at a level of technological sophistication where it is not clear that every technological advance can be seen to serve the interests of the public at large. We will discuss this topic in a little more detail in Chap. 8.

A simple example may illustrate that it is not always the most "technologically advanced" solution that is the best from a human perspective. Take the case of a humanoid robot given the task of caring for an elderly or infirm lady. On one level she will certainly appreciate help in having meals cooked, putting out the rubbish, hoovering, and the many other menial tasks which she may have difficulty performing on her own. She may even see her robot helper as a companion of sorts. However, tasks of a more intimate nature, such as helping her in and out of the bath, etc., may also be required. If the robot acquires, through advanced technology, a too humanlike appearance and/or demeanour (androidlike in our terminology), and this is an area where we may find that evolutionary techniques will be of great assistance, this may discomfit the elderly person, who may, for these functions prefer a more functional and matter-of-fact robot presence. Conversely, of course, there are other scenarios of an intimate nature where we may imagine that a more human and/or lifelike presence will lend itself to considerably greater user satisfaction.

Chapter 2
Evolutionary Algorithms and the Control of Systems

2.1 Control of Systems

Control of systems may reasonably be viewed as a branch of cybernetics as defined by Wiener: the "science of control and communication in the animal and the machine". One of its central subjects is that of control, and particularly feedback control of autonomous robots which is one of our core areas of concern. Indeed, when Wiener gave the new discipline of Cybernetics a name in 1948 he made use of the Greek word for steersman, *κυβερνήτης*; he arrived at this term through the etymology of the word "governor", a popular term used for the first widely used feedback device (Wiener 1948). Feedback control then is a topic of central importance in our understanding of biological and mechanical systems and in the design of robots and other machinery for the manipulation of our environment.

2.2 Recent Rapid Developments in the Field of Intelligent Robotics

The current rate of progress in the intelligent robotics field is quite astonishing and, if anything, appears to be happening at an ever-accelerating rate. Each new edition of the *IEEE Spectrum Robotics News*, an online publication of the Institute of Electrical and Electronics Engineers, showcases new developments in advanced robotics, many of which in their own way might be described as milestones in the field. For example, as of the time of writing, a robot (quadruped) has just been developed by the company Boston Dynamics (now owned by Google) which can run faster (albeit on a treadmill) than the world's currently fastest man.

Other recent developments described in this online publication include almost fully autonomous cars capable of navigating safely through urban streets

© The Author(s) 2015
M. Eaton, *Evolutionary Humanoid Robotics*,
SpringerBriefs in Intelligent Systems, DOI 10.1007/978-3-662-44599-0_2

(Google again), cooperating microrobots capable of building three-dimensional structures (SRI International), flying robots (hexrotor drones) capable of playing a symphony of musical instruments, including drums, bells, and a piano (KMel Robotics LLC), and a hopping bionic kangaroo (Festo AG & Co. KG). These are, of course, in addition to the recent developments in the specific field of humanoid robotics, which we address in Chap. 4.

2.3 Evolutionary Algorithms

Here we give a synopsis of the main evolutionary algorithms in current use in the evolutionary robotics field. We use the term "evolutionary algorithm" after Hoff-meister and Schwefel (1990), who used the term to cover algorithms which copied some principles from organic evolution and was meant specifically to refer to both the genetic algorithms of Holland (1975) and the evolution strategies of Rechenberg (1973). We use this term to cover the broader spectrum of any algorithms deriving in some fashion from the Darwinian evolutionary process. These include genetic algorithms (GA, Holland 1975), genetic programming (GP, Koza 1992), evolutionary strategies (ES, Rechenberg 1973), evolutionary programming (EP, Fogel et al. 1966), covariance matrix adaptation (CMA, Hansen and Ostermeier 2001), neuroevolution of augmenting technologies (NEAT, Stanley and Miikkulainen 2002), and the nondominated sorting genetic algorithm II (NSGA-II, Deb et al. 2002). As the genetic algorithm is probably the commonest evolutionary algorithm currently in use in the evolutionary robotics field, we devote the most space here to this paradigm. In fact, in our survey in Chap. 6 of the state of the art in the EHR domain, over 50 % of the research applications used some variant of a genetic algorithm as their main, or as an ancillary algorithm.

2.3.1 Genetic Algorithms (GA)

Genetic algorithms are search algorithms based on the mechanics of natural selection and natural genetics. They combine the notion of the survival of the fittest with a structured but randomised exchange of information between competing solutions. They also efficiently exploit historical information to improve performance over time. Genetic algorithms, or GAs as they are referred to in short, may be viewed as one of a family of algorithms operating around the same principles of natural selection and natural genetics, each with a different mode of implementation and each emphasising a different aspect of the natural process.

Genetic algorithms as proposed by Holland (1975) use three basic operators; these are reproduction, recombination or crossover, and, to a lesser extent, mutation. One starts with an initial population of structures, each of which encodes a specific solution to the problem at hand. This population is generally, though not necessarily, chosen at random. Each individual structure may take the form of a string of bits, or some other representational mechanism.

Of course, in a binary computer everything translates into bits at the end of the day; however, using, at a higher level, nonbinary representations (e.g., real values) the mutation and the crossover operators will have to operate in a slightly different fashion. Holland's original work demonstrated the ability of simple bit strings to encode complicated structures and also the power of simple transformations to improve dramatically the performance of these structures given sufficient time.

To deal with each of the basic operators in turn, mutation (random) provides background variation and occasionally introduces beneficial modifications into a structure. Mutation generally just involves changing a bit in the bit string from 0 to 1 or vice versa. Mutation is not assigned the same importance in genetic algorithms as in some of the other evolutionary algorithms, and so the probability of mutation is generally kept low. The crossover operation is generally looked on as the key to the power of the genetic algorithm. Crossover is probably best illustrated by a simple example (Fig. 2.1).

Assume these two strings encode different solutions to a particular problem ($x^1, \ldots, x^6$ and $y^1, \ldots, y^6$ taking either the values 0 or 1). If these two strings are selected for crossover the first thing to do is to select a crossover point or points. This crossover point is generally chosen at random, for a single-point crossover on a bit string of length l an integer position k is selected randomly in the interval $1, \ldots, l-1$. We can then create two new strings by swapping over all bits between the positions $k + 1$ and l inclusive.

Returning to our example, assuming a crossover point of 2, Fig. 2.2 shows how the strings (offspring) after crossover would look.

Consider two strings X and Y, each of length six bits:

X =	x^1	x^2	x^3	x^4	x^5	x^6

Y =	y^1	y^2	y^3	y^4	y^5	y^6

Fig. 2.1 Two simple chromosomes prior to crossover

X'=	x^1	x^2	y^3	y^4	y^5	y^6

Y'=	y^1	y^2	x^3	x^4	x^5	x^6

Fig. 2.2 Chromosomes after crossover

One entire bit string is sometimes known as a chromosome, with individual bits being known as genes. Crossover allows for genetic material to be passed from two chromosomes, which may be termed the parent chromosomes, to a create two new chromosomes, the offspring. This process allows the offspring to combine beneficial material from both parents. While some of the other evolutionary algorithms that we will discuss shortly emphasise mutation as the principal genetic operator, crossover is the main operator for genetic algorithms (Holland 1975). Without crossover, for an individual to acquire a beneficial trait requiring two separate mutations, neither of which on its own is beneficial, one of these mutations must happen to the parent, and then the second mutation must happen to one of that parent's offspring. This is an unlikely occurrence as the probability of survival of the first mutation will be low given that it will probably not have any immediate beneficial effect.

The final operator, reproduction, may be viewed simply as a process by which individual strings are copied according to their fitness. Very fit individuals receive a large number of copies; poor individuals possibly receive none. We start off with an initial population of strings, then, by the application of these basic genetic operators we hope to obtain populations increasing in overall utility. Summarising, a basic genetic algorithm to solve a particular problem will have the following components, many of which are shared with the other evolutionary algorithms which we will discuss shortly:

- a method of representing solutions to the problem in chromosomes
- a method of creating an initial population of chromosomes
- an evaluation or fitness function related directly to the problem environment for deciding the reproductive capability of individual chromosomes
- basic genetic operators for generating new solutions
- general parameters for the genetic algorithm such as population size, number of generations, etc.

Figure 2.3 gives an outline of a simple genetic algorithm (SGA) employing the so-called "roulette wheel" selection method, which simply involves the selection of an individual with a probability proportional to its fitness.

We are very conscious of the arguments put forward by Matarić and Cliff (among others) that if the amount of time and energy expended on the design of a GA for a particular problem, including tuning parameter sets and hand-crafting a fitness function for the particular problem domain and performing multiple experiments (necessary, given the inherently stochastic nature of evolutionary algorithms), exceeds the time required to hand-produce a particular control algorithm (or body design), then the use of evolutionary algorithms may not, indeed, be appropriate (Matarić and Cliff 1996).

With this caveat in mind, the author has been struck by the power of the simple GA to eke out solutions in complex problem domains. It may well be that using, for example, real-value encoding, tournament selection or some other selection procedure, or adaptive mutation or crossover operators will result in a more efficient search, but the power and beauty of the simple GA is that it works sufficiently well

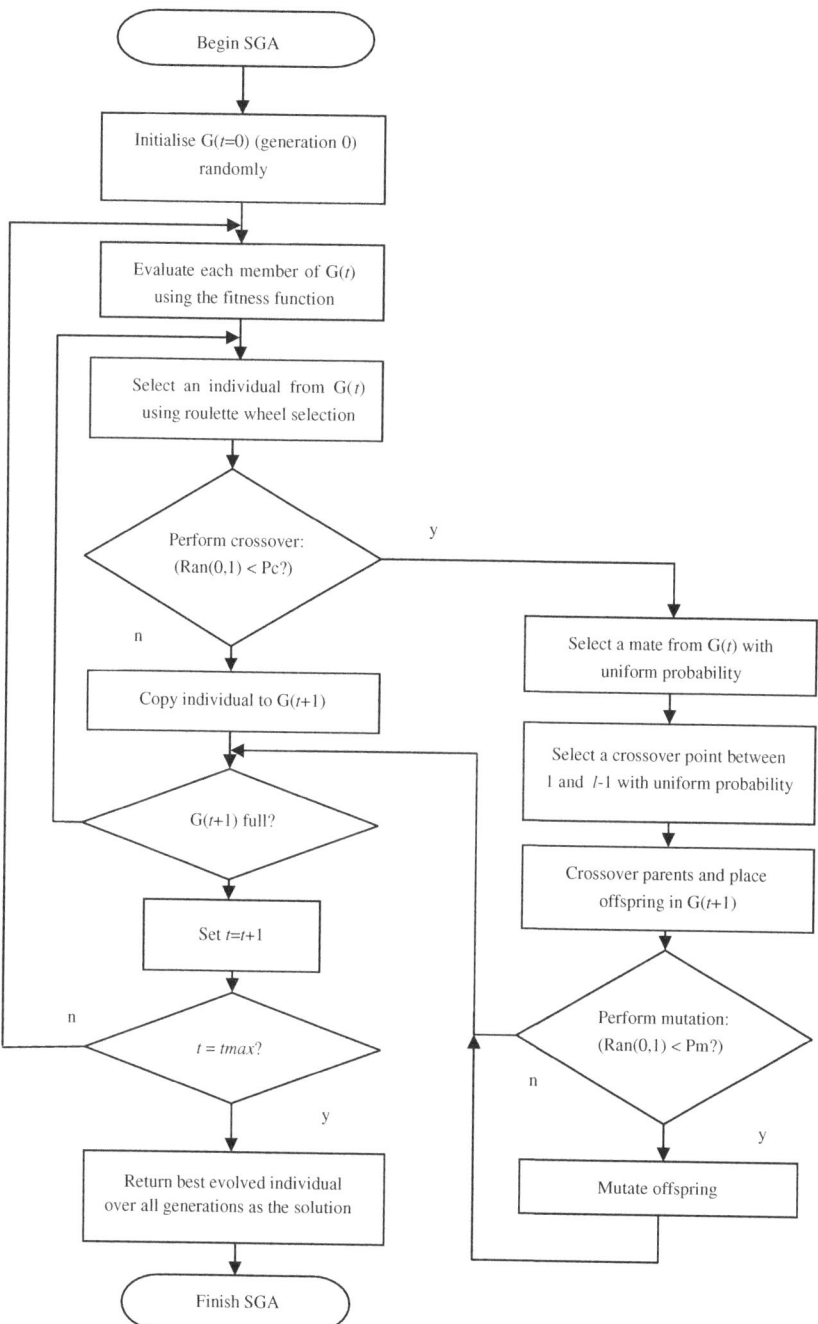

Fig. 2.3 Simple genetic algorithm (SGA) flowchart

in many circumstances. In many difficult problem domains, as in nature, we may not wish to find optimal solutions; indeed we may find it difficult to define exactly what it is we mean by optimal. In nature the core issues revolve around adaptation and survival of an organism in its environment, allowing for the transfer of an entity's genes from one generation to the next, or facilitating the transfer of a member of one's close genetic neighbors through to the next generation. In nature, the environment, to an extent, defines the fitness function.

Also, in nature the optimum shifts constantly in response to a changing environment, and perhaps in a more rapid fashion in response, for example, to the introduction of a new predator into a prey's environment. Indeed, in this case a finely tuned organism adapted specifically to a particular environment, and perhaps predator, may find itself at a loss (and probably quickly eaten!) when these changes occur. However, an organism with perhaps a more sloppy encoding system and with many redundant genes may find it easier to adapt through a system of neutral networks.

Our point here is that we see nothing sacrosanct about the SGA with a binary encoding mechanism, but rather that any evolutionary algorithm embedding the basic operators of mutation and/or crossover and selective reproduction has considerable power for the development of novel solutions to problems, especially over evolutionary time scales. These time scales may, of course, now be telescoped, using modern technology from thousands or millions of years into weeks, days, or even hours.

Kenneth De Jong, one of the foremost researchers in the field of evolutionary algorithms, puts it thus in his widely cited 2006 text *Evolutionary Computation: a Unified Approach* (De Jong 2006).

> ...asking what the goals and purpose of evolution are immediately raises long-debated issues of philosophy and religion which, though interesting, are ... beyond the scope of this book. What is clear, however, is that even a system as simple as EV appears to have considerable potential for use as the basis for designing interesting new algorithms that can search complex spaces, solve hard optimization problems, and are capable of adapting to changing environments.

The EV cited in this quotation refers to a simple evolutionary algorithm just employing the mutation operator (no crossover), not entirely dissimilar to the evolutionary algorithm used in the illustrative EA control example described at the end of this chapter.

Genetic algorithms are less commonly used on their own in the control of robots or other machinery, but instead are usually combined with some other (generally bioinspired) paradigm. By far and away the most common combination is to use a genetic or possibly some other evolutionary algorithm to evolve the weights and/or topology of a neural network-based controller. This neural network may be either of the feed forward or recurrent type. A recent example of this use of evolutionary algorithms to evolve neural networks of increasing complexity is the neuroevolution of augmenting technologies (NEAT) approach (Stanley and Miikkulainen 2002), which is outlined later in this chapter.

2.3.2 Genetic Programming (GP)

Unlike genetic algorithms, which in their most basic form operate on fixed-length binary strings, the genetic programming approach involves the evolution of programs of varying complexity. GP has been employed from the earliest days of research in the evolutionary robotics field. For example, in 1995 Gritz and Hahn used genetic programming for the generation of articulated figure motion, including the training of a simple simulated humanoid to perform a number of tasks, which included pointing and touching objects (Gritz and Hahn 1995). John Koza, the preeminent researcher in this field, has written several books on the subject, the first of which (a classic in its field) was published in 1992 (Koza 1992).

2.3.3 Evolutionary Strategies (ES)

The evolution strategy of Rechenberg (1973), described in (Hoffmeister and Bäck 1991), in its initial form used just mutation and selective reproduction. The probability of mutation, however, varies with time as stated in Rechenberg's 1/5 success rule to control the mutation parameter:

> The ratio of successful mutations to all mutations should be 1/5. If it is greater than 1/5 increase the variance, if it is less decrease the variance.

Rechenberg then extended the basic model to the so-called multimembered evolution strategy, which includes recombination. Mutation still, however, remained the main search mechanism. In addition, in an extension to the basic algorithm, the parameters of the system may also be subjected to change by the mutation and recombination operators. See Beyer and Schwefel (2002) for a detailed introduction to evolutionary strategies. Evolution strategies have been employed in several EHR applications, and form the core of the covariance matrix adaptation evolution strategy approach, described next.

2.3.4 Covariance Matrix Adaptation (CMA) Approach

The covariance matrix adaptation evolution strategy (CMA/ES) by Hansen and Ostermeier (2001), has formed the basis for a number of recent interesting research efforts in the evolution of both the morphology and controllers for humanoid and other legged creatures, mainly in simulation. CMA operates by adaptively varying the mutation distribution employed by the evolutionary strategy in order to make successful mutation steps that were made in the past more likely to occur in the future. Depending on the test function, speed improvements of several orders of magnitude have been observed when using the evolution strategy with covariance matrix adaptation, compared to using it without CMA. This evolutionary technique has been employed by several researchers recently in the EHR field mainly in simulation, including Al Borno et al. (2013) in the synthesis of a wide range of human movements, including walking and break dancing, Wang et al. (2012) for the synthesis of walking and running in a 30 degrees of freedom (DOF) simulated human adult, and Urieli et al. (2011) in the generation of soccer skills for a simulated Nao robot. A detailed description of this algorithm is beyond the scope of this text; the interested reader is referred to Hansen (2006) and Hansen and Ostermeier (2001) for a comprehensive description.

2.3.5 Neuroevolution of Augmenting Technologies (NEAT)

The neuroevolution of augmenting technologies (NEAT) approach was originally developed as an approach to solving complex control and sequential decision tasks and has in recent times has been used in the evolution of biped locomotion (Lehman and Stanley 2011; Allen and Faloutsos 2009a, b). NEAT works on the basis of starting the evolutionary process with a small number of relatively simple neural networks, and over time the complexity of the network topologies increases leading to the creation of separate species of networks as the number of generations increases. Further, steps are taken to ensure that a level of diversity is maintained as evolution progresses. Detailed discussion of the NEAT approach is outside the scope of this text; however, the interested reader is referred to Stanley and Miikkulainen (2002, 2004) and Stanley et al. (2005) for comprehensive overviews of the algorithm.

2.4 Evolutionary Algorithms for Control—A Simple Example

Here I describe one of my initial experiments with evolutionary algorithms, leading to control as an area of application. This work was initially stimulated by visiting the first International Joint Conference on Neural Networks (IJCNN) in Washington DC in 1989, where a poster presentation described a system using neural networks assumed to be contained in small "animals" moving in a two-dimensional world containing "food" elements (Cecconi and Parisi 1989).

In their system, the world in which the creatures reside is a 10 × 10 matrix of square cells, holding just a simple creature and a single food element. Four actions are possible by the creature at any time step: move forward one step, turn 90° right, turn 90° left, or do nothing. The creature receives information about the position of the food in terms of the distance of the food from the animal and the angle formed by the line connecting the creature and the food with some reference. Using the supervised learning back-propagation procedure, the creature was taught to predict the next position (angle, distance) of the food after each action.

In another part of the experiment the goal was to train the creature to approach and find the food. An 80 × 80 grid was used in this case, containing from 500 to 700 food elements. The creature was initially allowed to roam the environment selecting actions at random. Whenever it stopped on a food cell it was deemed to have "eaten" it and the food disappeared from the environment. Each time the creature stepped on a food cell the spontaneous activity was stopped. The creature recorded the sequences of actions that led it to the food cell and it was restarted in the position it had been in eight actions before, and these actions were repeated. This time, however, back-propagation was applied with the neural network receiving a training input on the output units coding the action selected by the animal. The training input was the action which had previously been selected at that point, as this had been a successful action leading the animal to the food. Cecconi and Parisi then went on to correlate ability at predicting the location of food to the ability to find and eat it. However, the section of most interest to me was at the end where they mentioned the case:

> in which the food approaching ability evolved in networks through genetic selection and random mutation

Details of these experiments were not provided, but my interest was by now sufficiently aroused to begin experimentation. My initial formulation concerned a simulated "beast no. 1" which would roam, as in the previous experiments, a two-dimensional world of squares containing a single food element. The size of the world was 16 × 16 squares, and 4 actions were possible, as in Cecconi and Parisi's experiments: to move one square in the direction being faced, to turn left, to turn right, or to do nothing. Sensory input to the beast was very simple, consisting of one of four inputs, either the food is forward (of the facing direction of the beast) to the left, to the right, or behind (Fig. 2.4).

The overall controlling program "Reality" called the two main routines, "Beast", for controlling the simulated creature, and "Genops" the genetic operators sub-routine controlling the production of new generations of "creatures" by selective reproduction ("Newgen" and "Select") and the application of the genetic operator of mutation. No other genetic operators were allowed, which was in keeping with the description of Cecconi and Parisi (1989).

Each beast was designed to operate as a stochastic automaton where the network weights represented the probability of transition from one state to the next. However, rather than updating the probabilities after each action (which assumes some sort of teacher or supervised learning, we assume that updating only takes place after a sequence of actions leading to either a successful (food is found) or unsuccessful (no food) outcome. As we are using an evolutionary algorithm to update the networks' weights, we are faced with the task of the encoding of the network structure in the chromosome. We chose a fixed chromosome structure of 32 bits arranged as in Fig. 2.5 (Eaton 1993b).

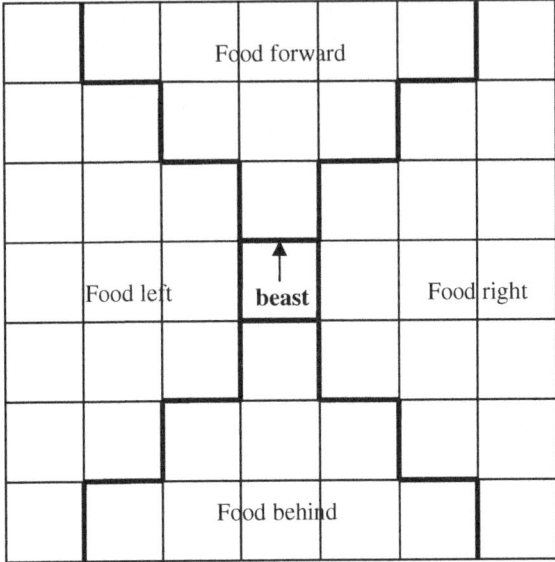

Fig. 2.4 Sensory input to "beast no. 1"

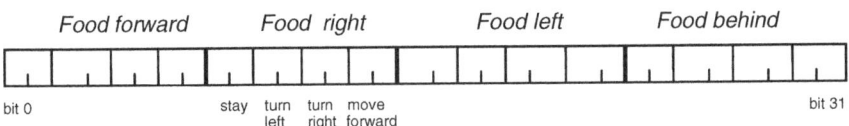

Fig. 2.5 Chromosome structure for simulated creature

In the initial experiments connections were first-order, i.e., the next action depended only on the present position of the food relative to the beast.

The chromosome encodes the probability of taking a particular action on the next time step based on a particular input. The probability of selection of individual actions were encoded two bits per weight using normal binary values which we then normalised so that the sum of the probabilities of action for a particular input was equal to one. As an example, assume that the food is currently ahead of the beast and the first eight bits of the chromosome are as in Fig. 2.6.

The probability of the beast remaining in the same position for this time step may be calculated as follows:

$$P(stay) = \frac{Val(stay)}{\sum Vals} \tag{2.1}$$

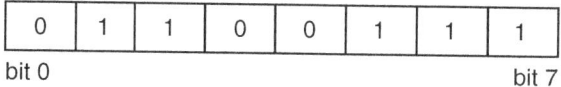

Fig. 2.6 Example chromosome section

where *Val(stay)* is the binary equivalent of the two-bit code in the corresponding locations in the chromosome, and $\sum Vals$ is the sum over all the values for this input. Returning to our example, the probability in this case of staying in position is 1/7 or 0.1428. If it happens that all of the bits are zero, an action is simply chosen randomly.

A fixed population of 30 controllers was allowed to run over a number of generations, starting from a random group of chromosomes, subject to the mutation operator. The mutation rate was allowed to vary over time to implement a form of annealing, and high mutation rates apply initially, which are reduced as the creatures converge on optima. Specifically, the probability of an individual bit being mutated was

$$P(mutation) = \frac{1}{fitness} \tag{2.2}$$

where the fitness of an individual was based on the number of successful runs over a number of trials combined with the number of steps per trial. Specifically:

$$fitness = 2^N \sum_{i=1}^{T} \frac{1}{C_i} \tag{2.3}$$

where N is the number of successful food runs in T trials, i.e., the number of times the beast successfully located the food with the beast and food being located at different randomly chosen locations for each run. C_i is cycle time per trial, i.e., the number of steps per trial. If the beast has not located the food after 10,000 time steps the trial is terminated, and this serves as the cycle time for this trial. The number of trials, T, was set equal to 10 for all simulations described here.

Once the system had been run, over typically 60 generations, the results were collected and graphs generated of maximum and average fitness per generation together with output showing the types of paths that the creatures took in locating (or not) the food. Interesting results were obtained; in general with the creatures learning to locate food early in the evolutionary process.

It is intriguing to note that the seemingly correct deterministic solution.

Algorithm FINDFOOD

While food-not-found
    If food-ahead **then**
          move-forward
    else if food-to-right **then**
          turn-right
    else if food-to-left **then**
          turn-left
    else turn-left **or** turn-right
    End If
End While

does not, in fact, work (interested readers might like to confirm this for themselves). The author found this, to his surprise, when he decided to upstage the evolutionary algorithm with some expert knowledge!

Chapter 3
Evolutionary Robotics (ER)

3.1 Introduction to Evolutionary Robotics

Evolutionary robotics (ER) involves the application of evolutionary techniques to the generation of either the "brain" (control systems) or to the "body" (morphology) of autonomous robots, or perhaps both. From a locomotive perspective, much research in the early days of research in the ER field (early and mid-1990s) involved the use of wheeled robots, in particular the ubiquitous Khephera robot. More recently, research has been conducted into evolving behaviours for legged robots, and the possible creation of robots with multiple means of locomotion (e.g., both wheeled and legged robots). As humans walk on two legs (bipedal locomotion) and as this text is concerned with the application of evolutionary techniques to humanoid robots, our main focus is on robots employing bipedal locomotion; however, it is also possible to envisage robots operating in a built-for-human (BFH) environment employing wheeled, or a combination of wheeled and legged, locomotion with the potential to adapt their mode of locomotion in response to differing types of terrain (Bongard 2013). These robots may not be overly human-like in appearance, however they will fall within our broad definition of humanoid. It should be noted that the well-publicised humanoid robot COG developed at the Massachusetts Institute of Technology in the early 1990s under Prof. Rodney Brooks, one of the foremost roboticists of our age, was a 21-DOF "upper-torso humanoid robot"; that is, it had arms, but no legs.

3.1.1 Categorisation of Research in ER in Terms of Level of Biological Inspiration and Level of Physical Realisation

We may categorise approaches to ER in two different fashions. First, we can categorise them in terms of the level of bio-inspiration; that is, the drawing of inspiration from biological processes, and/or attempting to mimic aspects of human/animal

© The Author(s) 2015
M. Eaton, *Evolutionary Humanoid Robotics*,
SpringerBriefs in Intelligent Systems, DOI 10.1007/978-3-662-44599-0_3

behaviour and/or morphology. Second, we can also categorise them in terms of physical realisation. Based on the second categorisation we can move from (a) simple control algorithms ("beast no. 1" as described in Chap. 2) to (b) simulated creatures with certain defined structures but that do not utilise realistic physics engines to generate their motions, to (c) those robots that observe a correct physical model (as those that can be simulated in Webots, Microsoft Robotics Studio, and other simulators with accurate physics), and finally to (d) robots with an actual physical realisation in a real embodied robot.

It should be noted that two (or more) of these levels of realisation may be combined and incorporated in a single experimental setup, e.g., it is a common practice to evolve control structures and/or robot morphologies inside a simulator with accurate physics, and then to transfer this evolved simulated system onto the real robot for verification and implementation.

3.2 Natural Versus Artificial Embodied Evolution

For many researchers ER translates into evolutionary motion design for (mobile) robots. However, from the perspective of natural evolution the environment the organism finds itself in (together with the imperative to survive and reproduce), *is* to an extent the fitness function, and an extremely complex fitness function at that. There is a question commonly asked by those skeptical of the power of natural evolution to evolve, without external intervention, complex organisms up to and including humans (and there are quite a few). This question relates to how evolution in nature manages to scale up from the production of simple organisms such as viruses and bacteria, to highly complex organisms such as the primate eye. A similar question currently arises in artificial evolution—how to move from the evolution of relatively simple behaviours (obstacle avoidance, wall following, etc.) to more complex ones. Any answers to this question found in the field of artificial evolution might provide pointers to solutions to the similar question as posed in natural evolution. Indeed, Eiben and Smith (2003, p. 263), in their book *Introduction to Evolutionary Computing* suggest that

> It could be argued that evolutionary robotics is the field where human engineers and scientists most closely approach natural evolution.

3.3 Fitness Functions for Evolutionary Robotics

A landmark article in recent years was by Nelson et al. (2009). This article, not uncritical of the field in general, surveyed the different fitness functions used over the years by a range of different authors, and evaluated and categorised these functions on the basis of the level of domain-specific knowledge contained within them. The next section is broadly based on the taxonomy as proposed by Nelson et al. with some minor additions/modifications

The fitness function specifies *what* task it is that we want the robot to perform, not *how* it is to perform that task. By evaluating the behaviour of the best-evolved robots for a particular fitness function we can judge the suitability of that fitness function in generating suitable behaviours, and, if necessary, modify the fitness function and again evaluate the evolved behaviours.

There is a certain element—at a high level—of interactive evolution happening here, an issue which we will come back to shortly.

Also, once we arrive at a fitness function that can generate high worth pheno-types, as evaluated by the closeness of the behaviours evolved to the actual desired behaviour (as opposed to high generated fitness values as generated by a particular fitness function), we may say that, in one sense, this fitness function encapsulates the core features of the desired behaviour.

3.3.1 Fitness Function Formulation

Nelson et al. (2009) conducted quite a comprehensive and far-ranging survey of fitness functions as employed in the evolutionary robotics field. Their research was based on looking at the degree of a priori information that was provided in the generation of different fitness functions. Their position was that the optimal fitness functions were those that provided the most novel control, while requiring the provision of least amount of domain-specific knowledge. They observed in their survey that the emphasis in much of ER work to date is concentrated on what might be considered relatively simple tasks, such as phototaxis (light seeking), obstacle avoidance, and, in the case of legged robots (bipedal or otherwise), basic loco-motion. They formulated a relative measure of the interest of a particular piece of research on the basis of the relative difficulty of the task evolved, taking into consideration the amount of a priori information supplied by the researcher. They pose the question as to whether the wealth of "proof-of-concept" research results, as published in the ER literature over the last 2 decades or so, can be transposed to more complex problem domains as might be important in humanoid robotics applications. They argue, in their influential article, that ER has as yet failed to evolve complex controllers for complex task domains, concentrating more on proof-of-concept applications. They also discuss the variety of tasks and behaviours that robots evolved using evolutionary techniques have been subjected to, including locomotion, object avoidance, gait learning, phototaxis, searching, foraging, and predator–prey task environments. A major part of their article involves the classification of fitness functions as used in the evolutionary robotics field into seven basic classes, based on the degree of a priori information that is incorporated into the fitness function design.

Their classification of fitness starts with training data fitness functions, which basically corresponds to a supervised learning scenario where a robot is trained to replicate behaviours displayed by a human or some other training agent. This classification ranges at the other end of the spectrum to aggregate fitness functions, which correspond in a certain sense to a reinforcement learning scenario, where only minimal feedback information is provided. This might correspond, for example, to the case of bipedal locomotion in a humanoid robot, using just the distance traveled (perhaps in the forward direction) before falling, or before a prespecified time limit. In between these two extremes they specify several other categories including behavioural, tailored, incremental, and competitive. Behavioural fitness functions, while not exactly saying what the expected response is to every input, do measure aspects of how the robot is acting (joint trajectories, etc.) rather than particular outcomes (distance travelled, etc.).

Tailored fitness functions, which form the majority of fitness functions surveyed in their study, and also in this book, combine aspects of both behavioural fitness functions and aggregate fitness functions. The authors give an example of such a function in a phototaxis task, where there could be an input to the fitness function which depends on the distance of the robot from the light source (aggregate function), but which also averages the distance the robot spends pointing in the direction of the light source (behavioural).

Nelson et al. identify two distinct forms of incremental evolutionary processes in their survey, using what they term functional incremental fitness functions and environmental incremental fitness functions. Evolutionary processes using the functional incremental fitness function approach are generally just referred to as "incremental evolution" or "staged evolution" in the literature. Incremental evolution simply involves the experimenter incrementally increasing the complexity of the evolved tasks as the evolutionary process proceeds, and this is a technique that has been employed by experimenters since the earliest days of research in the evolutionary robotics field. Both de Garis's and Lewis et al.'s early experiments in the evolutionary robotics field (which are discussed in our EHR prehistory section) used a form of incremental evolution (de Garis 1990a–c; Lewis et al. 1992). Of course, the use of incremental evolution requires some a priori knowledge on the part of the experimenter in deciding where to make the "breaks" in the evolutionary process, thus guiding the course of evolution to the extent that in a certain sense the evolved controllers might not be considered to exhibit truly novel behaviours (Nelson et al. 2009). While functional incremental evolution is quite a common approach taken by ER experimenters, the other form of incremental evolution described in this article, using environmental incremental fitness functions, is not so commonly used. This involves, rather than augmenting the complexity of the fitness function over the evolutionary process, instead increasing over time the difficulty and/or complexity of the environment in which the robots must operate.

This process of dividing a complex task into a group of simpler subtasks also has broad parallels in Brooks's subsumption architecture, where complex behaviours in organisms are generated by many simpler behaviours operating in parallel. In subsumption architecture layers (Brooks 1991) "are added incrementally, and newer layers may depend on earlier layers operating successfully, but do not call them as explicit subroutines".

A final type of fitness selection discussed in this paper is that of competitive and cocompetitive fitness selection. Of course, in a certain sense all forms of evolution involve a competition of sorts between individual members of a population for the right to reproduce and/or pass their genes on to later generations. However, competitive fitness functions involve direct competition between individual members of a population in the sense that the behaviour of one individual directly affects the others' behaviour and potentially the fitness that will be associated with those individuals. An interesting early example of competitive evolution was described by Sims (1994b), who describes an environment where two simulated creatures fight for control of a cube, each creature being encouraged by the evolutionary process not only to quickly approach the cube, but also to actively attempt to keep their opponent away. The winner in this instance was the creature closest to the cube after a set period of simulated time. Cocompetitive evolution involves two distinct populations performing distinct tasks competing against each other; a typical example here could be the coevolution of populations of predator and prey robots.

One type of fitness evaluation not discussed in detail in this survey is that involving interactive evolutionary computation (IEC). IEC involves replacing the objective function as used in the majority of EA applications in the robotics domain with a subjective evaluation of the performance of the robot based on the experimenter's opinion, or the opinion(s) of an independent observer, or group of observers.

This approach has been employed from the earliest days of research into the ER field including Lewis, Fagg and Solidum's pioneering work on the use of a GA in the control of a six-legged insect (which also incorporated a functional incremental approach) (Lewis et al. 1992). It is commonly employed when there is difficulty in constructing an objective fitness function due to the subjective nature of the problem domain, such as in the evaluation of dance routines, or in the appraisal of visual art or music.

3.3.2 Design of a Tailored Fitness Function

It could be said that the design of a tailored (or behavioural) fitness function is, in a sense, a form of interactive evolutionary computation, with the researcher him/herself forming the human component in evaluating the behaviour produced by the fitness function over a number of generations, and then tweaking the function in order to bring the observed behaviour closer to the "desired" outcome, whatever

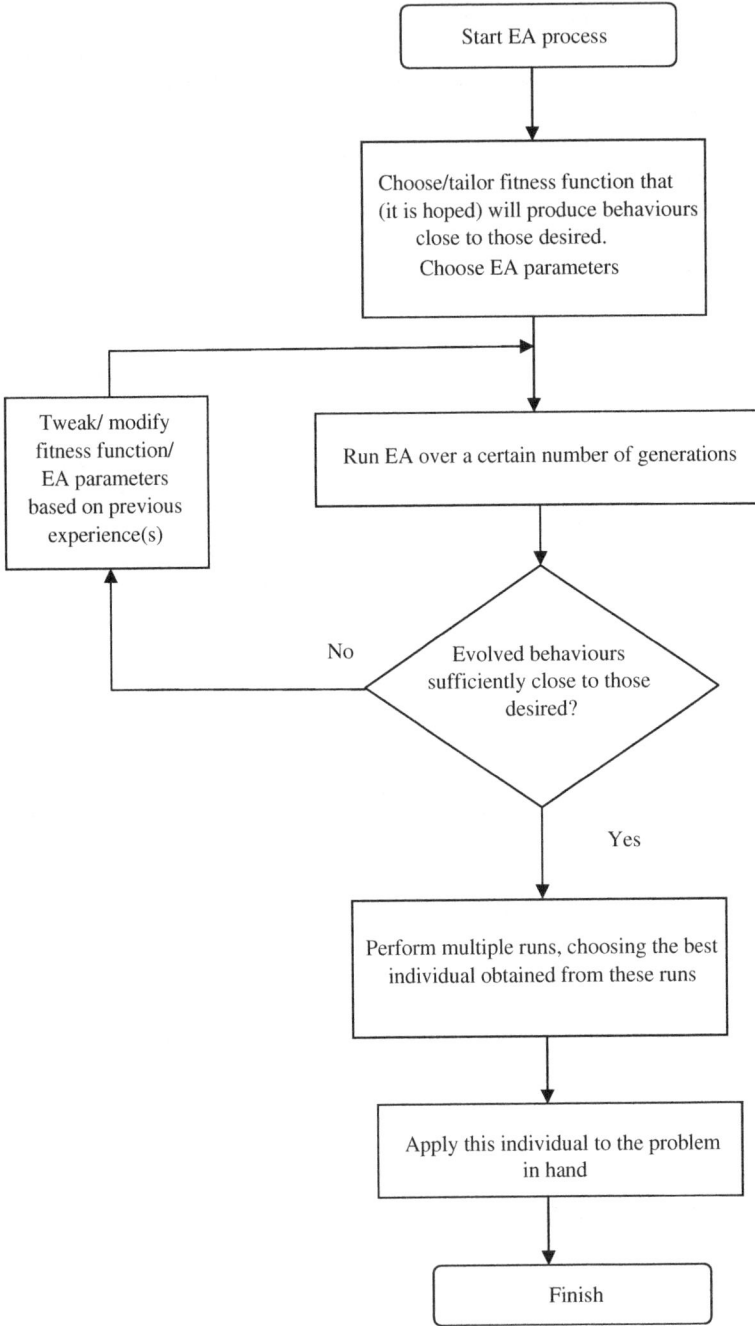

Fig. 3.1 Design of a tailored fitness function as a form of interactive evolutionary computation

this might be. Quoting from Nelson et al. (2009) relating to behavioural and tailored fitness functions

> These types of fitness functions are formulated by trial and error based on the human designer's expertise.

See Fig. 3.1 for an outline flowchart illustrating this general idea. Of course, other aspects of the evolutionary process may also be chosen on this empirical basis, such as crossover and mutation probability, choice of genetic operators, etc.

To recap, the notion of the creation, by an iterative process, of a fitness function itself is, in a sense, a form of interactive evolutionary computation. We note that, of course, the idea of having a "meta" evolutionary algorithm explicitly operating at a higher level in order to choose factors such as mutation and crossover rates, etc., is not new. For example, in Grefenstette (1986) the author used a metalevel GA to choose six separate parameters for a lower-level GA, including mutation and crossover rates, population size, and whether an elitist or nonelitist survival strategy was employed.

3.4 Evolving "Interesting" Behaviours—What Exactly Constitutes an Interesting Behaviour?

Much work in the ER community has been involved in the evolution of "interesting" behaviours. We would like to briefly explore what it is that makes a behaviour interesting in the eye of the researcher, and the level of domain-specific, or a priori knowledge or information which the researcher supplies as related to the "interestingness" the evolved behaviour exhibits. As Nelson et al. (2009, abstract) put it,

> The underlying motivation… is to identify methods that allow the development of the greatest degree of novel control, while requiring the minimum amount of a priori task knowledge from the designer.

For example, in the field of evolutionary humanoid robotics (EHR), where the goal is to evolve aspects of the behaviour and/or morphology of a robot which has human-like characteristics, if we use a bipedal robot with two arms, one evolved behaviour could consist of the robot flailing around wildly on the floor, with no apparent purpose. To most observers this would be considered an uninteresting behaviour. On the other hand, the robot could behave in a highly predictable fashion, performing a single simple movement in a repetitive fashion, for example, raising one arm and then lowering it repeatedly. Again, to most observers this would constitute an uninteresting behaviour. However, if the evolutionary process produced another movement, again rhythmic, however, this time moving one foot, and then the other in sequence, bending its knees, and moving its arms in such a fashion that it avoids falling over—then this behaviour would be considered interesting to most observers—"Look it can walk!" (or dance, or whatever).

So, in quantifying what we mean by "interesting behaviour", we need to explore the region between random unorganised movement on one hand, and "boring" repetitive behaviour on the other.

3.5 Why Is ER Still a "Fringe" Topic for Many Robotics Researchers?

It is probably fair to say that the EHR field is still in a stage of relative infancy—the topic of evolutionary robotics is a little more than 2 decades old, and the first articles addressing the application of evolutionary techniques to recognisably humanoid robots in particular appeared not much more than a decade ago. It may, however, be legitimately argued that 2 decades (for the evolutionary robotics field in general), while a short period by the standards of most disciplines, is still a substantial length of time in terms of the still relatively youthful domain of robotics in general, and certainly in the more specific area of autonomous mobile robots. Why, then, does this subject remain a "fringe" topic for many researchers in the field, and why in many university robotics curricula is the topic of evolutionary robotics consigned very much to the edge of the curriculum? Perhaps there is perceived by some to be a certain sense of fragmentation and disillusionment within the evolutionary robotics community, from the early heyday of Floreano, Cliff, and Sims' (and many others) work in the early 1990s, where it seemed anything (and everything!) might be possible. A certain sense of disenchantment with the field seems to have crept in where one is almost reluctant to say precisely what one's field of research is at international robotics conferences for fear of not being taken too seriously. As Stanley (2011) points out, evolutionary robotics might not yet be considered a mainstream topic in robotics.

The field of evolutionary robotics in the early 1990s might be compared to a certain extent to the fledgling Artificial Intelligence (AI) movement in the early 1950s in that very great things were promised, with possible huge benefits to humankind. In the now famous Dartmouth summer research project (McCarthy et al. 1955) on AI it was proposed to conduct a 2-month study on

> how to make machines use language, form abstractions and concepts, solve kinds of problems now reserved for humans, and improve themselves

It was considered that "significant advances can (could) be made in one or more of these problems" if a "carefully selected group of scientists" worked together for a 2-month period!

The pioneering computer scientist Alan Turing was no less optimistic in his landmark publication "Computing Machinery and Intelligence", published 5 years earlier (Turing 1950). Turing believed that

at the end of the century [i.e. the year 2000] the use of words and general educated opinion
will have altered so much that one will be able to speak of machines thinking without
expecting to be contradicted

Strong stuff indeed. And while great advances have indeed been made in various
fields of AI, including computer vision, natural language recognition, object classification, and so on, this has been over a period of some 60 years, not the 2-month
period rather optimistically forecast by John McCarthy, Marvin Minsky, and others
at the Dartmouth summer school. And while great advances have also been made in
areas such as expert systems and automatic reasoning, very few of us even today
would say that computers yet have the ability to think.

And so this is, perhaps, to a certain extent the situation in the evolutionary
robotics field. Much was expected, which by and large has not yet been delivered to
date. And although the field is still relatively young (about 20 years old) compared
to the AI field, perhaps it is now worth briefly looking at some of the areas where it
has delivered results, and some where it has not lived up to early expectations. We
will also look at some of the reasons behind these failures and how these obstacles
might be overcome, and finally perhaps take a pragmatic look into the future and to
the contribution of evolutionary robotics to the development of the increasingly
complex and sophisticated mobile robots of the future.

3.5.1 Four Critical Areas for Success in the Application of EHR Techniques

There are four specific areas which we feel will be critical in the success of the
application of evolutionary techniques to humanoid robots and to robots in general.
These are the issues of:

(1)   The level of incorporation of a priori fitness function knowledge of the problem
       domain in question. In some instances, the creation of carefully crafted and
       complex tailored fitness functions, honed to a specific problem domain, may
       involve a similar level of difficulty to that of directly hand-coding a solution to
       the problem in hand. There is thus a perception that many fitness functions are
       overly finely tuned—why not just put all of this effort into hard-coding the
       behaviours instead? As Mataric and Cliff (1996) succinctly put it

For a real reduction in human effort, the effort expended in designing or configuring the
evolutionary system should be less than that required to manually design or configure the
robot controllers that it produces.

Also, to quote from Nelson et al. in their comprehensive survey of fitness
functions for evolutionary robotics applications (Nelson et al. 2009)

It was found that much of the research made use of fitness functions that were selective for solutions that the researchers had envisioned before the initiation of the evolutionary processes. The degree to which features of evolved solutions reflected a priori knowledge on the parts of the human researchers varied.

(2) Crossing the "reality gap" from simulation to embodied robotic implementation where behaviours evolved in simulation are not guaranteed a successful transition to the real robot.

Because of the potential for damage to both the robot and to its environment involved in evolving behaviours on real robots (especially true in the case of attempting to evolve from scratch behaviours on an adult-size humanoid robot), together with the time scales involved, some or all of the evolutionary process is generally consigned to simulation. However, if the simulator used does not faithfully represent the robot and its environment, difficulties arise in the transfer of evolved behaviours from simulation to the real robot—this disparity of behaviours is termed the reality gap. We discuss this issue further in Chap. 4.

(3) Overcoming the "scalability barrier", which involves using evolutionary techniques in the generation of increasingly more complex behaviours, rather than the relatively simple "proof-of-concept" behaviours favoured by many researchers to date. Relating to the perceived lack of ability of solutions obtained from the ER community to "scale up" to real-world problem domains consisting of multiple goals and complex task sets, Nelson et al. (2009) suggest

The fundamental question of how to select for truly complex intelligent autonomous behaviors during evolution remains largely unanswered and evolutionary robotics remains somewhat on the fringes of autonomous robotics research.

(4) The issue of benchmarking of results, given the wide variety of different simulators, robotic platforms, and tasks under consideration.

Doncieux et al. (2011) put forward a cogent set of arguments in this regard in their recent article "Evolutionary robotics: exploring new horizons". They argue that much of the work done in the evolutionary robotics field to date has been concerned with proof-of-concept type experiments, to determine the feasibility of a particular evolutionary approach to a robotics problem domain, be it simulated or embodied. However, they suggest that, in many cases, authors fail to build on previously reported experiments, preferring to create new systems from the ground up. This, of course, can also lead to issues such as problems with the verifiability of results and with the evaluation and benchmarking of different experimental results, in order to create controllers and/or robot morphologies of increasing utility. This is a topic we return to in more detail in Chap. 7. They suggest that to mature as a discipline "ER need[s] less proofs of concepts and more solid results".

Summarising then, some of the currently perceived failings/shortcomings in the ER field are:

- the lack of concrete results, outside of a limited range of problem domains
- the lack of scalability of results
- the reality gap issue
- the lack of appropriate benchmarking of results
- the production of controllers whose performance and robustness cannot be assured.

3.6 How Might Perceived Failings in the ER Field Be Addressed?

All of these issues are being addressed, in one form or another, by members of the ER community. For example, the reality gap issue in particular has been mitigated to some extent, in recent times, by the introduction of flexible simulators with accurate physics engines capable of reproducing, with high accuracy, complex evolved behaviours. However, this, on its own, may not be enough; a number of researchers are actively looking at ways to reduce this gap through a variety of ingenious techniques. We present a synopsis of some research results in this area in the next chapter.

Regarding scalability of results, a number of researchers are actively working on solutions that will both scale up to real-world problem domains, and that will not require carefully crafted fitness functions in the process.

The issue of performance evaluation and benchmarking is of significant importance in order for the field evolutionary robotics, and autonomous mobile robotics in general, to move forward. We consider this topic to be of sufficient importance to devote all of Chap. 7 to it.

In general, perhaps we need to have a certain sense of pragmatism in realising that, for the creation of real-world systems, evolution on its own may not be enough: this approach may not be the solution to all problems relating to the design of controllers, sensors, and morphology of autonomous mobile robots. This involves a realisation that it may be necessary to sensibly combine our research results with robotics practitioners of a more conventional bent. We note here that the issue of benchmarking in particular is one which affects not just evolutionary robotics researchers, but many other researchers in the autonomous mobile robot research field.

It is this author's view that, in the future, evolutionary techniques, maybe not on their own, but in conjunction with (and perhaps even forming the gel that binds together) several other AI methodologies (both symbol-based and non-symbol-based), will form a core component in the construction of future lifelike humanoid robots of great utility to mankind.

In conclusion, it seems clear that evolutionary techniques will play a significant role in the development of future humanoid robots. Exactly the scale of that contribution is not yet clear; however, there exists a clear proof of concept in the overarching power of the natural evolutionary process: humanity itself.

Chapter 4
Humanoid Robots, Their Simulators, and the Reality Gap

4.1 Introduction to Humanoid Robotics

The field of humanoid robotics is concerned with the creation of robots which are broadly humanlike in their behaviour, their morphology, or both. The definition of what constitutes a humanoid robot is quite broad in the literature, so in this chapter we suggest our own abbreviated taxonomy. The idea of the creation of a being in mankind's own image harks back into the realms of prehistory, one could even say back to the Genesis creation narrative, where a new creature named woman "because she was taken out of Man" was created from the rib of Adam.

More recently there has also been the trend, with advancing technology, to create startlingly humanlike humanoid robots, thus triggering the so-called uncanny valley effect (Mori 1970). If, however, we define a humanoid robot in a broader sense as a robot which is simply designed to operate in environments designed for human inhabitants (our level "built for human" or BFH), then this robot may, in fact, look nothing like a real human.

Although there has been an explosion in published literature and research related to humanoid robotics in the last few years, the vast majority of this research has been "nonevolutionary" in nature, so the task of writing this book is not as daunting as it might seem at first sight. What has been more difficult has been delving into the main strands of disparate research that lead to the field as it stands today.

© The Author(s) 2015
M. Eaton, *Evolutionary Humanoid Robotics*,
SpringerBriefs in Intelligent Systems, DOI 10.1007/978-3-662-44599-0_4

4.1.1 Some Potential Applications of Humanoid Robots

We outline here a few of the areas in which humanoid robots may find application in future years; these include:

- areas in which humans normally operate which may involve unpleasant, dangerous or tedious work (the so-called 3 Ds—dirty dangerous, or dull), or some combination of these three types of unattractive labour
- home help
- care of the young, the elderly, and/or the infirm
- entertainment applications (in a broad sense)

These are some of the early application areas where humanoid robots may be seen to replace and/or augment human labour. In the first category, these robots may be generally seen as a boon, relieving humans of unpleasant or dangerous tasks. Of course, in developing economies such robots might be seen as a threat to the livelihood of many workers on the margins of society, who are barely eking out a living, such as those scavenging for items on waste dumps. However, it is unlikely that it would be economically viable to use expensive humanoid robots for such a task, so such livelihoods, such as they are, are safe. In the third category also (including care of the elderly and/or the infirm) humanoid robots are also generally likely to receive a warm welcome, certainly in developed economies such as Japan and Italy, with rapidly ageing populations. As regards home helps, humanoid robots may well be seen, in the not so distant future, as useful tools, just like the washing machine or the Hoover, which it is likely that most households of reasonable income will aspire to own. Unlike in the earlier part of the 20th century, most households except the most affluent do not have servants, so it is unlikely that much human labour will be displaced. Of course, we could also take the example of poorly paid "day labourers", who may operate on a cash basis. In this instance, humanoid robots, of a reasonable cost, might indeed threaten livelihoods. In the final category (entertainment robots) it is probably likely that, certainly in the initial stages, humanoid robots will augment rather than displace humans.

However the picture overall is not entirely rosy. It is interesting that Alan Turing, in his 1951 essay "Intelligent Machinery, A Heretical Theory" [published posthumously, and reproduced in Copeland (2004)], saw the possibility of great objections to the construction of such machines from two main sources. One source he identified as coming from religious groups. The other source he identified was coming from "intellectuals who would be afraid of being put out of a job". Looking to the future, the first wave of "intelligent" humanoid robots is very unlikely to pose a major challenge to academics, teachers, engineers, physicians, and anybody who holds a position of moderately complex intellectual or physical nature, or anybody operating in a variety of environments (door-to-door salesman, marketing executives, etc.). However, Turing concludes this essay with the rather chilling comment

At some stage therefore we should have to expect the machines to take control …

Hopefully this will not transpire to be the case!

4.1.2 Our Criterion for Inclusion as a Humanoid

Many of the robots discussed in the next section (e.g., the Nao or the Bioloid humanoids), which are currently used in EHR research, would not actually qualify as "humanoid" if our criterion is the level n BFH—that is, that they are capable of operating in "built for human environments". They are simply too small.

However, given that their morphologies are similar, in a broad sense, to those of humans, we would hope that the behaviours evolved in these robots would transfer without too much effort to a "full-sized" (child- or adult-sized) humanoid. This is an issue which we will touch on again later in this chapter, in our discussion on techniques to bridge the reality gap.

Looking through the different experiments conducted in applying evolutionary techniques to humanoid robots over the years, one thing becomes clear: the wide variety of robotic platforms and their associated simulators which are used by researchers. If any "standard" platform/simulator can be said to have emerged in recent years, it is that of the Nao humanoid robot and its associated Spark simulator, probably due to its adoption as the RoboCup standard platform league robot in 2008, replacing the previous incumbent, the Sony Aibo quadrupedal robot.

4.2 Our "Definition" of Humanoid (Levels of Anthropomorphicity)

As it is possible to describe many different robots that would fall within the generally accepted category of humanoid, which, however, would vary wildly in intellectual and physical capabilities, here we attempt to outline a brief taxonomy, to differentiate between robots with different characteristics. Note that this taxonomy should not be seen as set in stone, instead it could be viewed as a useful starting point in this regard. See Bar-Cohen et al. for another discussion on the topic, together with a list of terms commonly used to describe robots with humanlike features (Bar-Cohen et al. 2009).

We note here that, inherent in our discussions of the different types of humanoid robots, we only include robots which incorporate the notions of embodiment and locomotive abilities of some sort. These restrictions exclude robots which exist purely in simulated form such as robots involved in the RoboCup simulated humanoid robot league, and robots such as the MIT COG project of the late 20th century, which involved the construction of a robot whose upper body was broadly humanoid, and had significant cognitive capabilities, but which did not possess locomotive ability (wheels or legs). This is not to say that these robots should not fall within a more general categorisation of humanoid robots, just that they do not fit within this particular taxonomy.

Indeed, about half of the research projects discussed in the next chapter on the current research projects in the EHR field involve robots that exist in simulation alone. Here, then, is our outline taxonomy for $n + 1$ separate levels of embodied humanoid.

4.2.1 Level 0: Replicant

Identical to humans in every physical and behavioural aspect, except for obvious differences in the eating of food, drinking liquids, passing of bodily wastes, etc.

Would not normally be able to distinguish from a human, using normal human senses. Could be termed "replicant level".

4.2.2 Level 1: Android

Not quite at the replicant level, but very close in every aspect to human morphology and behaviour. Very high levels of intelligence and dexterity. Could be termed "android level".

...here we may skip some level(s) depending on the details of the final taxonomy employed...

4.2.3 Level n−3: Humanoid

Close to a human in both "body" and "brain", however, there is no possibility of mistaking the robot for a human being except, perhaps, at far range. High levels of intelligence and dexterity. Could be termed "normal humanoid level", or just "humanoid level".

4.2.4 Level n−2: IH (Inferior Humanoid)

Could not be mistaken for a human, however, has the broad morphology of humans—2 arms, 2 legs, bipedal, stereo vision, and auditory facilities. Reasonable intelligence and dexterity however may be confined mainly to a limited task set. Could be termed "inferior humanoid level", or "IH level".

4.2.5 Level n−1: HI (Human-Inspired)

Looks quite unlike a human, however, has the broad morphology of humans, has either bipedal or multipedal capabilities; may also be wheeled. Limited intelligence and dexterity, generally confined to a limited task set. Could be termed "human-inspired level", or "HI level".

4.2.6 Level n: BFH (Built-for-Human)

Looks nothing like a human, however is able to operate in most environments designed for human inhabitants. Generally designed to capably perform a limited task set. As these robots are designed to operate in built-for-human environments, we could term this the "built-for-human robot level" or "BFH level" [after Brooks et al. (2004)].

4.3 Brief Overview of Selected Humanoid Robot Platforms

4.3.1 Introduction

In this section we provide an overview of some of the humanoid robot platforms currently in use, and of those humanoid robots which have been used over the last decade in the evolutionary robotics field. Many of the platforms used, especially in the early years in the application of evolutionary techniques to humanoid robots, were custom platforms, particular to the research being conducted; we do not look at these here, instead we refer the reader to the original publications. Because of the large and ever-growing number of humanoid robots under development and currently in use, we will not have space to discuss these in depth. The reader is referred to recent survey articles, see Akhtaruzzaman and Shafie (2010) and Duran and Thill (2012) among others for a more detailed discussion of humanoid robot platforms in general.

4.3.2 Humanoid Robot Platforms Used in Evolutionary Robotics Experiments

Probably the commonest humanoid robot platform currently used in EHR experiments is the Nao robot from Aldebaran robotics (Fig. 4.1). A variety of behaviours have been evolved using this robot, including omnidirectional walking, dance, ball kicking (for RoboCup applications), and dance choreography. Another commonly

Fig. 4.1 Nao humanoid
(Domingues et al. 2011)

used platform is the Bioloid humanoid robot (assembled from a kit) from Robotis,
Inc., Korea (Fig. 4.2). This robot has been used as the basis for the development of
walking in a variety of environmental conditions, headstands, standing up, and in
the evolution of dance using noninteractive evolutionary techniques. The Humanoid for Open Architecture Platform (HOAP) series of robots from Fujitsu, Japan,
has been used in a variety of ER experiments including walking, the generation of
cooperative dance and kicking motions, and for the evolution of yoga motions.

Among the "full size" humanoid robots employed are the Exciting Nova on
Network (enon) wheeled robot, also from Fujitsu (see Fig. 4.3), the Dexter "adult
size" humanoid from Anybots, Inc. USA, and the iCub humanoid, designed to be
the same size as a 3-year-old child (Fig. 4.4). Table 4.1 summarises some of the
important characteristics of recent humanoid robots which have been used as the

Fig. 4.2 Bioloid humanoid.
Photo taken by the author

basis for evolutionary robotics experiments, either embodied, or in simulation. Examples of evolved applications for each of these humanoids are also given, the reader is referred to Chaps. 5 and 6 for further details of these various applications.

4.3.3 Other Humanoid Robot Platforms

As we have just mentioned, there is such a wide variety of humanoid robot platforms either in use or under development today that we will just discuss briefly here what we see as the most important platforms. None of these robots have, to this author's knowledge, yet been used in an evolutionary robotics context. This does not, of course, confirm that this is the case, since research may be ongoing which has not yet been submitted for publication. There also exists the possibility, given the possible huge potential commercial applications of these robots, that research is being conducted which is not intended for publication, for commercially sensitive reasons. There is also undoubtedly research being conducted into the potential military applications of autonomous humanoid robots for battlefield applications; whether any of this research is evolutionary in scope is outside this author's sphere of knowledge. However given the potential advantage that could be gained by

Fig. 4.3 enon humanoid
(Uchiyama et al. 2011)

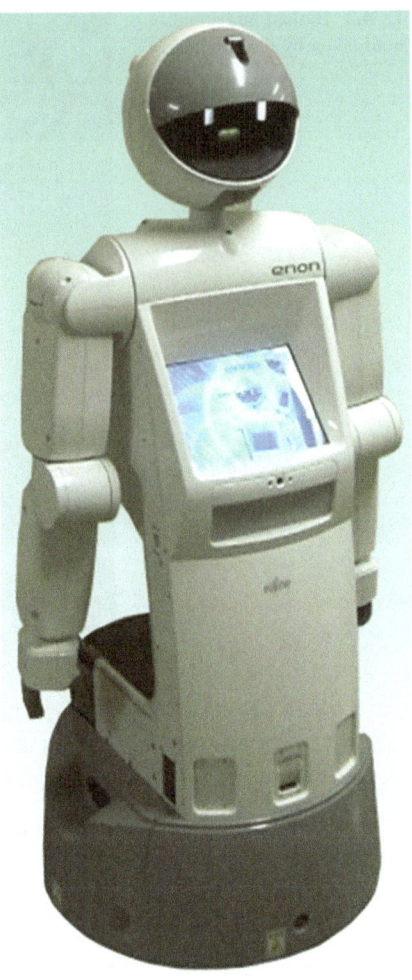

battlefield robots to be able to autonomously adapt and change to evolving hostile environments, it seems unlikely that some research of this nature is not already under way.

4.3.3.1 ASIMO

Since its first introduction in 2000, and following many years of research and development in the field of humanoid robotics, starting with the E series in the 1980s and early 1990s, it is probably fair to say that, in many people's eyes, Honda's ASIMO (standing for Advanced Step in Innovative Mobility) remains the archetypal humanoid robot. ASIMO can both walk and run (at speeds of up to

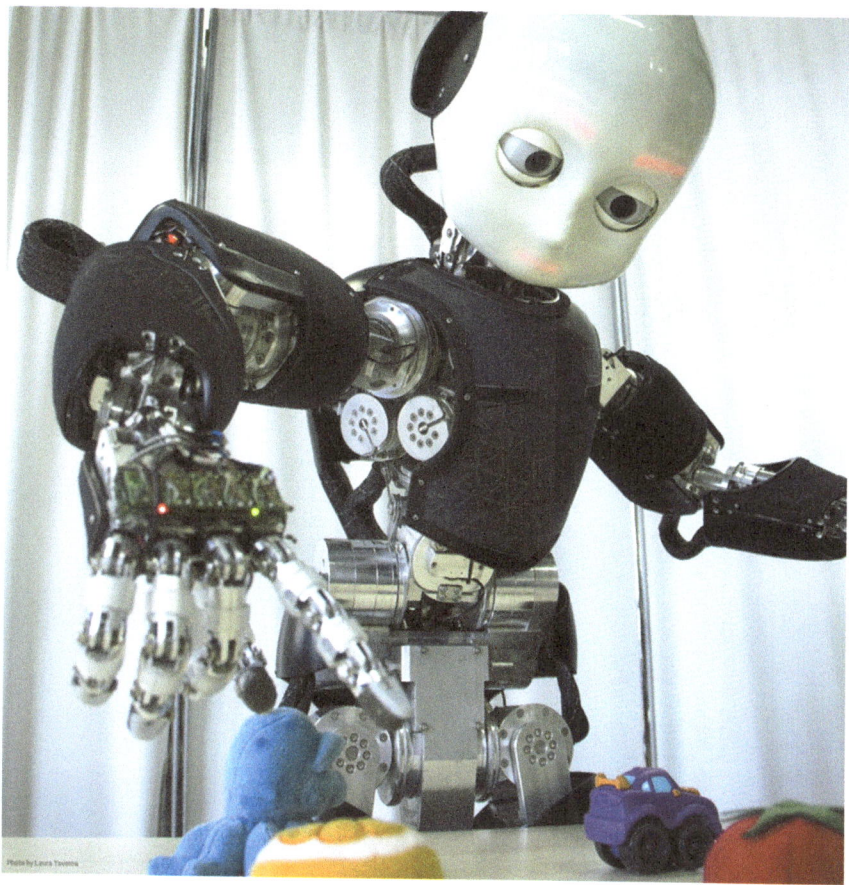

Fig. 4.4 iCub robot. http://www.iit.it/en/social/photo-gallery.html. Accessed 9 September 2014.
Thanks to Giorgio Metta of iCub.org for permission to reproduce this photo

6 km/h), and the current version incorporates a total of 57 degrees of freedom.
ASIMO can also walk up and down stairs, and perform dance routines. ASIMO
weighs 48 kg and its height is 130 cm, which Honda determined was the ideal
height for a robot that would be able to act as a mobility assistant, with the
capability of turning on and off switches, opening and closing doors, serving food
and beverages, etc. ASIMO has made numerous public appearances which include
in 2008 conducting the Detroit Symphony Orchestra, and in 2002 ringing the bell at
the New York Stock Exchange.

In response to the 2011 Great East Japan Earthquake, Honda has also been
accelerating research into the possible adaptation of ASIMO to work effectively in
extreme hostile environments such as those encountered in the aftermath of such a
disaster.

Table 4.1 Humanoid robot platforms used in evolutionary robotics

Manufacturer/developer	Humanoid platform	Degrees of freedom (DOF)	Example of evolved application(s)	Dimensions: height	Weight (kg)
Anybots Inc., USA	Dexter	12-DOF	Generation of bipedal locomotion (in simulation only); generation of standing balance for the robot in both simulation and in reality	An 'adult-size' humanoid robot	Not specified
Korea Institute of Science and Technology (KIST), Korea	Humanoid robot 'MAHRU'	35-DOF	Generation of ball-catching type movement	150 cm	67
Fujitsu, Japan	Frontech enon (Exciting Nova on Network) service robot, wheeled	16-DOF in total (Head: 2-DOF, Drive wheels: 2-DOF, Arms: 5-DOF per arm, Hands, 1-DOF per hand)	Performing as an exhibition guide robot	130 cm	Approximately 50
Yamagata University, Japan	Bonten-Maru I	23-DOF	Gait synthesis, stair climbing	120 cm	32
EU project RobotCub	iCub humanoid robot	53-DOF in total	Reaching and grasping tasks, modelling those of human infants, applied to a 14-DOF model of the iCub humanoid robot (simulation only)	105 cm	24
Aldebaran Robotics, France	Nao humanoid robot	Various: 14, 18 and 25 degrees of freedom. Specialised 21-DOF model used for RoboCup competition	Dance generation; omni-directional walking for RoboCup applications; humanoid robot walking, comparing three different learning algorithms; evolution of dance choreography	58 cm (Nao NextGen)	4.3

(continued)

Table 4.1 (continued)

Manufacturer/ developer	Humanoid platform	Degrees of freedom (DOF)	Example of evolved application(s)	Dimensions: height	Weight (kg)
KAIST Robot Intelligence Technology (RIT) Lab, Korea	HSR-VIII	26-DOF	Footstep planning for a humanoid robot in the presence of obstacles (in simulation only)	53 cm	5.5
Fujitsu, Japan	HOAP-2 (Humanoid for Open Architecture Platform) robot	25-DOF	Evolution of 2 asymmetric yoga motions, 11 and 14 joints (DOF), respectively	50 cm	7
Fujitsu, Japan	HOAP-1 robot	20-DOF	Generation of various motions for humanoid robots including cooperative dance and kicking using IEC. Optimise motions generated using conventional GA	48 cm	5.9
Robotis Inc., Korea	Robotis Bioloid humanoid	18-DOF	Generation of bipedal locomotion for normal, low-friction, and reduced-gravity conditions; investigation of the effect of morphological constraints on movement; generation of a variety of behaviours including performing a headstand, walking, and standing up, based on initial human-supplied postures; dance generation	40 cm (Type A humanoid)	1.7
Kondo, Japan	KHR-2HV	17-DOF	Generation of humanoid robot movements —walking	34 cm	1.3
Kondo, Japan	KHR-1	17-DOF	Forward walking	34 cm	1.2

4.3.3.2 HRP-4C

The HRP-4C humanoid robot is a 158-cm-tall humanoid robot developed by the Japanese National Institute of Advanced Industrial Science and Technology (AIST). HRP-4C has a realistic female head and face ("android" level) and has been designed to have the bodily proportions typical of the average young Japanese female. AIST, in collaboration with Kawada Industries, Inc., developed a series of humanoid robots, including the 154-cm-tall HRP-2 in the period 1998–2002, and the 160-cm-tall HRP-3 developed in 2002–2006. Work then commenced on the development of a "cybernetic human", HRP-4C, over a 2-year period using mechanisms adapted from the HRP-2 and HRP-3 robots. In the words of Kaneko et al. (2009)

> The word "Cybernetic Human" is a coinage for us to explain a humanoid robot with a realistic head and a realistic figure of a human being.

The main initial design goals for the HRP-4C robot were

(A) Capability of bipedal walking
(B) Realistic figure of a human being
(C) Configuration to imitate humanlike motion

while design goal (B) was adapted after some consideration to be

(B) Realistic figure of the average young Japanese female

HRP-4C has 42 DOF in total and weighs 43 kg and is capable of humanlike movement, including bipedal locomotion and dancing. Recent hardware improvements have increased the obvious potential applications of HRP-4C in the entertainment industry (Kaneko et al. 2011).

The robot S-One from SCHAFT, Inc., a company operated by group of researchers who left Tokyo University to develop this robot, is based on the HRP-2 humanoid robot. This robot recently won first place in the DARPA Robotics Challenge trials scoring a total of 27 points, out of a potential maximum of 32 points.

4.3.3.3 HUBO2/HUBO2++/KHR-4

HUBO2 is the latest in the HUBO series of robots which has origins in the year 2000 with the KHR-0 robot, which was a bipedal robot without arms or an upper body, developed in a machine control laboratory in KAIST (Korea Advanced Institute of Science and Technology). This research was inspired, in part, by Honda's newly unveiled ASIMO. The KHR-1 robot followed in 2003 and KHR-2, which was fully humanoid in shape, with a head and functioning hands, followed in 2004. The KHR-3, then titled HUBO, was developed and in 2005, in order to celebrate the 100th anniversary of the announcement of the theory of special relativity. KAIST and Hanson Robotics joined forces to create Albert HUBO, a variant of the KHR-3, with the addition of a realistic animatronic head modeled on the famous scientist (Oh et al. 2006). This added head allowed for the generation of a variety of expressions emulating surprise, happiness, sadness, etc.

The HUBO2 robot, also known as KHR-4 was developed in 2009, and has a total of 40 DOF. It has a height of 1.25 m, weighs 45 kg, and is capable of walking at 1.5 km/h and running at 3.6 km/h. HUBO2++ is a variant of HUBO2, taking into account issues of "users' convenience" (Heo et al. 2012).

Variants of HUBO2 were entered by two teams for the United States Defense Advanced Research Projects Agency (DARPA) robotics challenge, team DRC-HUBO based at Drexel University, and Team KAIST.

4.3.3.4 Atlas/PETMAN

One of the most impressive humanoid robots in existence at the present time is the Atlas humanoid robot, developed by the American company Boston Dynamics, now owned by Google. This robot was based on the PETMAN robot, said to be designed for testing chemical protective clothing, whose prototype (PetProto) in turn was derived from existing the existing BigDog robot (Raibert et al. 2008). These robots were developed using funding and oversight from DARPA. Atlas stands approximately 1.8 m tall, and weighs 150 kg, using a construction based on titanium and aircraft-grade aluminium.

As of the time of writing at least six of the teams competing in the 2015 finals of the DARPA Robotics Challenge (DRC) will use robots based on the Atlas humanoid; all of these teams will, in addition qualify for DARPA funding. The lead organisations behind these teams are the Florida Institute for Human and Machine Cognition, the Massachusetts Institute of Technology, TRACLabs, Inc., a collaboration between Worcester Polytechnic Institute and Carnegie Mellon University, Lockheed Martin Advance Technology Labs and a joint collaboration between TU Darmstadt Germany, TORC Robotics, Virginia, USA, Oregon State University, and Virginia Tech.

Atlas has 28 degrees of freedom, and is capable of independent operation, however, it is often operated tethered to an external power supply, which also serves to maintain stability. The PETMAN humanoid, on which Atlas is based, was designed with a body shape closely conforming to a 50th percentile male body shape (Nelson et al. 2012).

4.3.4 Conclusion

This concludes our brief survey of the state of the art in humanoid robots, both those which have been the subject of evolutionary experimentation and also some notable current humanoids. Of course, this discussion is in no way comprehensive, and the reader is directed to the survey articles mentioned earlier and to manufacturers' own literature and websites. It should be noted, however, that because of the secretive nature of the research conducted in some quarters (for commercial and other reasons) it may be difficult in some cases to obtain the most up-to-date information on particular robot platforms. For example, Honda succeeded in keeping their initial research in this area completely secret for over 10 years with only top management being informed; no papers were published and research was conducted in a room without windows (Menzel and D'Aluisio 2000).

In conclusion, it is only fair to mention that humanoid robotics, as a separate research field, has its origins in the WABOT (WAseda roBOT) series of robots initiated in Waseda University in Japan in the 1970s. Things have moved on quite a bit since then!

4.4 Simulators Used for the Evolution of Humanoid Robots

4.4.1 Introduction

In this section we briefly review some of the simulators which can be used in the generation of behaviours for autonomous mobile robots, and humanoid robots in particular. As Pinciroli et al. point out (Pinciroli et al. 2012), it is only in the last 10 years or so that general-purpose simulator environments have become available to the ordinary researcher, due to advances in computing technology based mainly on CPU speed and multicore architectures, and vastly increased memory capabilities. These flexible platforms allow researchers the freedom to create new robot models and simulated environments with a fraction of the effort than was previously required in setting up tools from scratch for each new experiment. Many of these modern simulators employ the open-source Open Dynamics Engine (ODE) physics library at their core.

Obviously, the more accurate the robot simulator employed, the higher the chances are of the successful transferral of evolved behaviours to the real robot, an issue we will look at shortly in our discussion of the "reality gap". This term refers to the discrepancy in performance between behaviours evolved in simulation and the actual performance of the real robot, once these behaviours have been transferred over from the simulator.

A wide variety of software platforms are in existence today, with others currently under development. Indeed as Sato and colleagues succinctly suggest (Sato et al. 2008)

> the aptly named YARP, i.e., Yet Another Robot Platform, describes the situation of the software platforms well.

Our interest mainly focuses on those simulators demonstrating flexibility in terms of robots and environments which can be modeled, and which can perform accurate modeling of the robot's behaviour in three dimensions, Some earlier simulators only operated in two dimensions, these, while useful for modeling wheeled robots such as the Khephera robot, are obviously of limited applicability for the modeling and simulation of complex humanoid robots. Some of the simulators discussed here follow the open-source model, while some are commercial products. Some simulators employed by researchers use custom software, which is not freely available; these fall outside the scope of our discussion.

4.4.1.1 Open Dynamics Engine (ODE)

The Open Dynamics Engine (ODE) which was developed by Russell Smith and several other contributors, is a library for the simulation of rigid body dynamics. It is suitable for simulating both wheeled and legged robots, so is suitable for use in the field of evolving humanoid robots, and comes with collision detection built in. ODE places an emphasis on rapid and stable simulation over physical accuracy of simulation, which ensures robust operation. While ODE comes with a built-in graphics front end, it is typically used in conjunction with a higher level environment, such as Webots or Gazebo. For a general survey and review of several different physics engines, including ODE, PhysX, Newton Physics Engine, and the Bullet Physics Library, see Boeing and Bräunl (2007).

4.4.1.2 Webots

The most widely used simulator in recent years in the application of evolutionary techniques to humanoid robot design appears to be the Webots simulator from Cyberbotics (Michel 2004). This is a commercial simulator whose development started in 1996 at the École Polytechnique Fédérale de Lausanne (EPFL) under the direction of Olivier Michel. The company, Cyberbotics Ltd., was then founded in 1998 as a spin-off venture. As such, then, this is one of the one of the oldest simulators available today that has been under continuous development from its inception, and which allows for the modeling of 3D robots using an accurate physics engine. According to recent publicity from the company, the software is in use at the time of writing by over a thousand universities and research centers. Webots uses the ODE for its dynamics simulation.

A variety of robots are available for simulation under Webots, both wheeled or legged, or the user has the option to build their own, as we did in creating the Bioloid humanoid model for use in our evolutionary humanoid robotics experiments. Once the behaviour has been created/evolved in simulation this can then be transferred to the real robot. Our early experiments in the evolution of bipedal locomotion were initially conducted on a QRIO-like simulated robot supplied with the Webots package (Eaton and Davitt 2006). We then developed a crude model of the Bioloid humanoid robot which allowed for the evolution in simulation of bipedal locomotion. Our current experiments involve the use of an accurate model of this humanoid that we have developed over the last number of years, which allows for the evolution in simulation of behaviours including bipedal locomotion and dance, and the subsequent transfer of these behaviours into the real robot with relatively small discrepancies of behaviours between simulation and reality (Eaton 2007b, 2013).

A number of researchers have used Webots to simulate the Aldebaran Nao humanoid robot for a variety of tasks, mainly in the area of applications related to the RoboCup Standard Platform League, for which the Nao is the robot employed. For example, Kulk and Welsh used a simulated Nao robot to evolve an improved walking performance for the real Nao robot, and compared their results with those obtained using two other algorithms (Kulk and Welsh 2011). Webots has also been used to evolve behaviours for the HOAP-2 humanoid (Hettiarachchi and Iba 2010) and (in simulation only) for the 26 DOF KAIST humanoid robot HSR-VIII (Hong et al. 2009).

4.4.1.3 Open Architecture Human-Centered Robotics Platform (OpenHRP)

OpenHRP consists of both a simulation environment and a motion-control library for application to humanoid robot control. Collision detection is incorporated in the simulator, and the consistency between humanoid behaviour in simulation and on the real robot is emphasised (Kanehiro et al. 2004). Yanase and Iba used this simulator for the generation of a variety of motions for humanoid robots, including dance and kicking motions (Yanase and Iba 2006, 2008a). The latest incarnation of this simulator (OpenHRP3) is claimed to be a considerable improvement on the previous versions (Cisneros et al. 2012).

4.4.1.4 USARSim

USARSim, standing variously for Urban Search and Rescue Simulation (Carpin et al. 2007), or for Unified System for Automation and Robot Simulation (Balakirsky and Kootbally 2012), is a simulator based on the Unreal Development Kit (UDK) produced by Epic Games, which was originally devised as a development kit for the creation of first-person shooter-type games. For an early example of work using the Unreal game engine using evolutionary algorithms for dynamic path selection in an immersive 3D environment (set on the University of Limerick campus) see Eaton et al. (2002). USARSim was initially developed by the US National Science Foundation (NSF) for use in urban search and rescue scenarios including the RoboCup Virtual Robot Rescue competition; however, to date it has been used in a variety of robot simulation contexts including the DARPA Urban Challenge and the IEEE Virtual Manufacturing and Automation Challenge (Balakirsky and Kootbally 2012). USARSim employs the PhysX physics engine for accurate modeling of real-world physics. Van Noort and Visser recently used USARSim in the validation of the dynamics of the Nao humanoid robot (van Noort and Visser 2012). In the specific context of evolutionary humanoid robotics Antonelli and colleagues used the

USARSim simulator in the simulation of the Kondo KHR-2HV humanoid robot, which has 17 degrees of freedom. Ten parameters were evolved by the robot, nine representing joint angles, the tenth is a time-frame value. Using the USARSim simulator in a semi-interactive basis for the first 50 generations, walking behaviour was developed for this robot which was then transferred to the real robot for the final stages of evolution (Antonelli et al. 2009).

4.4.1.5 Gazebo

Gazebo is a simulator originally designed to accommodate multirobot simulations in a three-dimensional outdoor environment. It was conceived towards the end of 2002 at the University of Southern California by Andrew Howard and Nate Koenig as a simulator to augment the abilities of Stage, an existing simulator designed for 2D simulations of interior environments. Gazebo is available under the open-source license and, in common with several of the simulators described here, uses ODE as its core physics engine (Koenig and Howard 2004). Gazebo's current main claim to fame is probably its adaptation by DARPA in April of 2012 as its official simulator for the Robotics Challenge project; this is discussed further in Chap. 8 on benchmarking issues. It is claimed that this selection was based on an "informal market survey" of currently available simulators.

While Gazebo has an august history (and potential future) in the field of simulation environments, there appears to be little evidence to date of its employment in evolutionary humanoid robotics applications.

4.4.1.6 SimSpark

The SimSpark simulator is a general-purpose simulator based on the ODE that allows for the simulation of general-purpose multiagent environments. The project started in 2003 and has its origins in the Spark multiagent simulator, which was used as the first official simulator for the RoboCup simulation league in three-dimensions (Obst and Rollmann 2005). As such, it is probably best known today for its use in the simulation of teams of Nao humanoid robots for the current RoboCup 3D soccer simulation league.

SimSpark has also been used specifically in the evolution of behaviours for humanoid robots, in particular for the Nao humanoid robot (as currently used in the RoboCup Standard Platform League). Urieli and colleagues conducted research using the SimSpark simulator in evolving robot soccer skills (in simulation only) using the CMA/ES (covariance-matrix adaptation evolutionary strategy) including kicking, walking, and turning (Ureili et al. 2011). In 2011 Domingues et al. also used the SimSpark simulator to generate walking behaviour based on partial Fourier series optimised using a genetic algorithm, again for the Aldebaran Robotics Nao humanoid robot. In this case the evolved behaviours were successfully transferred onto the real Nao robot, while the authors acknowledged that some modifications were needed in the transfer from the SimSpark simulator to the real Nao robot (Domingues et al. 2011). Picado and colleagues in 2009 also applied genetic algorithms for the offline evolution of gait parameters for walking gaits for the Nao robot in simulation only (Picado et al. 2009).

4.4.1.7 Other Simulators

There is a wide variety of simulators available on the market today with the potential for modeling humanoid robots to generate and test behaviours in simulation by evolutionary or other means. Space precludes a detailed discussion of these simulators, many of the recent ones have been designed to efficiently simulate multiple robot swarms, and some of which are more focused on 3D game development. These include Autonomous Robots GO Swarming (ARGOS) (Pinciroli et al. 2012), Microsoft Robotics Developer Studio (MRDS) (Jackson 2007), SwarmSimX (SSX) (Lächele et al. 2012), Modular Open Robots Simulation Engine (MORSE) (Echeverria et al. 2011), RealitySim (Fu et al. 2011), Multi-robot-simulation Framework (MuRoSimF) (Friedmann et al. 2008), Panda3D (http://www.panda3d. org), and the popular long-running low fidelity first-order motion simulator, Stage (Vaughan 2008).

Simulators aimed the simulation of particular humanoid robots are also available for use. One of the most notable of these is the iCub humanoid robot simulator, developed as part of a European project with the aim of creating a new open-source robot platform for cognitive robotics research. The initial simulator was developed using the Webots package described earlier; however, the current iteration is a standalone open-source platform (Tikhanoff et al. 2008, 2011).

4.4.2 Conclusion and Summary

If one is prepared to pay what is admittedly a fairly substantial upfront premium (Webots PRO, allowing access to the full range of Webots features, currently retails from around CHF 2300; the pared down EDU version at the time of writing is a relative bargain at CHF 320), then the Webots simulator would appear to be the simulator of choice.

Of the open-source options the Gazebo simulator has one of the longest pedigrees, however, it does not, to date, appear to have had much application in the EHR field. USARSim and OpenHRP are also viable alternatives in this regard. For evolutionary experimentation involving the Aldebaran robotics Nao robot the SimSpark simulator would appear to currently be the simulator of choice. However, many researchers still prefer the route of creating their own tailored simulation environments, typically utilising the ODE library to provide the core physics functionality.

4.5 Crossing the "Reality Gap" Between Simulation and Embodied Robots

4.5.1 Introduction

The "reality gap" is an issue which has engaged the interest of many researchers in the ER and EHR fields. Of course, if one is in possession of a simulator which can perfectly replicate the conditions in which the humanoid will operate, together with the morphology and the control mechanisms of the robot, then this gap vanishes. Unfortunately, no such replicator/simulator currently exists.

Some approaches to the reality gap issue involve quite complex operations, involving for example, modeling the fitness landscape, and the modification of the simulator; we will discuss here some of these approaches. However, we suspect that, in the long-term future, as simulators become more accurate, it may be more productive to put extra effort into building more faithful models of the robot(s) and its environment, than into highly complex "avoidance" strategies.

4.5.2 The Reality Gap, and Approaches to Minimise Its Impact

We will now look at steps the can be taken, and that are being used by researchers to minimise the impact of the reality gap; that is, the discrepancy between behaviours evolved in simulation and the actual observed behaviours on the real robot, resulting in inefficient controllers. It has been suggested that there is a trade-off between the efficiency of simulation in an evolutionary robotics context, and in the likelihood of a faithful translation of these evolved behaviours to the real robot (humanoid or otherwise). This suggests that efficient behaviours in simulation may exploit inaccuracies in the simulator which exploit phenomena which have little or no bearing in reality (Koos et al. 2013).

The reality gap is, of course, an issue in any circumstances when we wish to initially generate a behaviour in simulation and then to transfer this behaviour, or set of behaviours (evolved or otherwise), to a real robot. One possibility, is to avoid the use of a simulator altogether, and to create motions directly on the actual robot itself by evolutionary means or otherwise. A difficulty, however, then arises when we seek to create behaviours which may cause strain on the robots actuators, or may indeed be physically impossible. This issue is more pertinent to the evolutionary robotics field than to many other areas because of the stochastic nature of EAs. As a result of this randomness it will be possible, indeed likely, especially in the early stages of evolution, to generate motions that could be damaging to the robot, its operator, or the environment in which it is situated (or, indeed, a combination of all three). The possibility of this occurrence may be reduced by "seeding" the initial population, that is, by generating an initial population of controllers which we know are not going to cause problems for the robot or its environment. However, this requires a degree of a priori knowledge about what a "correct" controller should look like, and we typically wish to minimise this input of domain-specific information.

This situation would be particularly serious in the case of attempting to evolve behaviours directly on a full-size humanoid robot, such as the "Cybernetic Human" HRP-4C (Kaneko et al. 2009), given the expensive nature of the hardware involved, together with the potential for damage. To this author's knowledge no evolutionary experimentation has been done to date to evolve complex behaviours directly on a full size humanoid robot.

Of course, the "holy grail" of this approach would be to have a robot (or group of robots) which, as Alan Turing described in his far-seeing work "Intelligent Machinery" published in 1948 and reproduced in Copeland (2004), would

> be allowed to roam the countryside

and

> should have a chance of finding things out for itself

However as Turing acknowledged:

> the danger to the ordinary citizen would be serious

We note also that the potential for damage in a limbed/legged robot is considerably higher than that for a wheeled robot, given morphological considerations. For these reasons it is likely that, at least for the foreseeable future, at least part of the evolutionary process for humanoid robots of any moderate size will take place in simulation.

Interestingly, as pointed out in (Koos et al. 2013), even if evolution is carried out completely on the real robot without resource to a simulator, because of the necessary constraints placed on the robot in the experimental environment, the evolved controllers may not transfer well to real environments in which the robot will be expected to operate.

We will now look in turn at some of the approaches researchers have taken to attempt to minimize the effect of this gap on the utility of evolved robots in real-world environments. Most of the approaches discussed are proposed by researchers in the evolutionary robotics field, however, the reality gap is also an issue for any researchers attempting to create, by any means, behaviours for subsequent transfer to the real world. As the main focus of this section is to give a broad overview of some of the different approaches taken by researchers to the reality gap issue, we will not be overly concerned with details of the evolutionary algorithms employed, fitness functions, etc. For a further recent overviews of approaches to the reality gap issue see Koos et al. (2013) and Zagal and Ruiz-del-Solar (2007).

4.5.2.1 The Transferability Approach (Koos et al. 2013)

A recent approach to tackling the reality gap issue is the transferability approach (Koos et al. 2010, 2013). This approach is based on the observation that, as briefly discussed earlier, there is experimental evidence to suggest that there is direct inverse correlation between the efficiency of solutions evolved in simulation, and the possibility of their successful implementation on a real robot. This hypothesis led to a two-pronged evolutionary approach which is applied to both the fitness and also the estimated transferability of the results obtained by the simulator. This transferability is measured by a simulation-to-reality (STR) measure which looks at the disparity between simulated and actual behaviours. Their results were validated using an 8-DOF wheeled-legged (WHEG) quadrupedal robot on a walking task, together with a navigation task using an e-puck wheeled robot (Mondada et al. 2009) in a T-maze environment. The robot used in the walking task is constructed from a Bioloid kit, as used in our experiments for investigating the evolution of bipedal locomotion and dance in humanoid robots (Eaton 2008a, 2013). This experiment corresponded broadly to that performed by Jakobi in his early experiments investigating the reality gap issue, where the Khephera robot used by Jakobi was replaced by the e-puck robot. In their experiments the transferability approach was shown to clearly outperform Jakobi's envelope of noise approach for the e-puck task, and successful walking robots were also evolved for the quadrupedal robot, which were found to transfer well to reality.

4.5.2.2 The Grounded Simulated Learning (GSL) Approach (Farchy et al. 2013)

The grounded simulation learning approach involves the modification of an imperfect simulator in order to make it correspond more closely to reality. It is an iterative approach involving initially evolving a behaviour(s) in simulation, then taking this evolved behaviour and implementing it on the real robot. Any discrepancies between the behaviour on the real robot, compared to that evolved in simulation are noted, and the simulator is then modified in order to make it conform closer to reality. This process is then repeated until a satisfactory correspondence between the behaviour(s) evolved in the simulator and those observed on the real robot are achieved. The robot platform used was the Aldebaran Nao humanoid robot which was applied to a bipedal locomotion task. Starting from an initial set of hand-coded parameters, a substantial increase in walking speed of was observed over four iterations of the GSL approach. The simulator used is the SimSpark multi-agent simulator as used in the RoboCup 3D simulation league, however because of the difficulty involved in altering the simulator itself, instead the inputs provided to the simulator were altered. A certain amount of expert guidance was also provided as it was found that, following the first iteration, altering some of the parameters resulted in unstable behaviours so these parameters were removed from subsequent iterations.

4.5.2.3 Leveraging Multiple Simulators Approach
(Boeing and Bräunl 2012)

In this approach the researchers suggest the use of multiple simulators in the evolution of control strategies. Evolving across multiple simulators results in different fitness values being applied to controllers evolved in different simulation environments. These individual fitnesses are then combined using statistical methods to come up with an overall fitness measure for the controller, and the evolution progresses. It is suggested that the use of multiple simulators will have a similar effect to the addition of noise to the robot dynamics, and that the process of evolving on multiple simulators provides variance analogous to the transfer of controllers from simulation to reality, thus providing additional robustness to evolved designs. The simulators used were the Ageia PhysX dynamics simulator, with and without Smoothed Particle Hydrodynamics (SPH), and the Newton Game Dynamics simulator accessed via the Physics Abstraction Layer (PAL) software package, and applied to the control of the "Mako" Autonomous Underwater Vehicle (Boeing and Bräunl 2012). They conclude that the use of multiple simulators in itself does not guarantee evolved solutions better than those evolved on a single simulator, however, this approach does appear to improve the likelihood of evolved designs successfully crossing the reality gap.

4.5.2.4 Combining Evolution in Simulation with Preprogrammed
Behaviours (Duarte et al. 2012)

In this approach a set of preprogrammed behaviours, known to transfer well from simulation to reality, are used to bootstrap the evolutionary process. The idea here is to combine the ability of evolutionary algorithms to create novel behaviours with a set of handpicked primitive actions known to be core to the particular application domain. The robot used in these experiments is the e-puck robot (Mondada et al. 2009), and the application domain is a double T-maze simulated using JBotEvolver (an open-source simulation engine). The primitives used in this work consist of the simple preprogrammed behaviours follow-wall, turn-left, and turn-right. The authors acknowledge, however, that the choice of these handpicked primitives may restrict the search space to suboptimal solutions, but suggest the possibility of using a hierarchy of synthesised behaviours in an attempt to address this issue.

4.5.2.5 Using an EA to Tune the Parameters of a Simulator (Laue
and Hebbel 2009)

The objective of this approach is to use an evolutionary algorithm (in this case an evolution strategy) in a multistage approach in order to determine the parameters for an existing simulation engine which will result in the closest possible approximation of the behaviour in the simulator to its behaviour in the real world. The robot used in this work was the (discontinued) Sony AIBO robot, and 38 parameters in all were optimised. These comprise three PID control parameters for each of six motors together with the maximum velocity and torque for each motor, six friction parameters as well as two global parameters used by the SimRobot simulator, which is the simulator used in this work. The application was in the accurate simulation of a variety of different walks on the AIBO robot, including suboptimal walks. In order to evaluate the accuracy of the simulation, the results were compared with the well-respected Webots simulation environment. The quality of walks using the initial (untuned) parameters was generally inferior to the Webots model; however, after the first stage of learning the simulation with the optimised parameters outperformed the Webots model.

4.5.2.6 Fitness Function Correction Interleaving Simulated and Real Data (Iocchi et al. 2007)

This approach involves the interleaving of experiments on the simulator and on the real robot in the ratio 1:5 (one experiment on the real robot to every 5 in simulation). The purpose of the interleaving process is to determine discrepancies between the results for the simulator as opposed to those from the real robot; these discrepancies are then used to alter the fitness function which evaluates results from the simulator (as opposed to the simulator itself). An advantage of this approach is that it is easier to modify a function than it is to modify a simulator. The task in this case is to implement walking and kicking behaviours in a 22-DOF humanoid robot, Robovie-M. The simulator used in these experiments was USARSim.

4.5.2.7 Coevolution of Controller and Simulator Using Estimation–Exploration (Lipson et al. 2006; Bongard and Lipson 2004)

This approach involves a coevolutionary process in which the controller and the simulator are evolved in an iterative fashion. Starting off with a basic simulator (which may not approximate the actual behaviour of the robot very well), we evolve the desired behaviour(s); this is termed the exploration phase. The best controller is then taken and implemented on the real robot. Because of inaccuracies in the simulation this is unlikely to produce the performance desired, especially if this is the first iteration and we are starting with a fairly crude simulator.

The simulator is then itself evolved so that it reproduces the behaviour observed in the real robot correctly, based on the currently optimal controller. This is the estimation phase. We now take the best-evolved simulator and use this to evolve a new controller, and the cycle repeats until a controller is found which exhibits satisfactory performance on the real robot. The target robot used was a quadrupedal robot; in the estimation phase the masses each of the nine body parts of the robot together with the time lags for each of eight sensors were evolved.

4.5.2.8 The "Back to Reality" Approach (Zagal et al. 2004; Zagal and Ruiz-Del-Solar 2007)

Similar to the estimation–exploration algorithm just discussed, the "back to reality approach" involves the coevolution of both the robot controller (and/or its morphology) and its simulator. The authors argue, however, that their approach is more applicable to real robot experiments, as the feedback used to alter the simulator is based just on differences in observed behavioural fitnesses as opposed to sensor data comparisons in which small differences between the simulated and real robot can lead to uncorrelated signals (Zagal and Ruiz-del-Solar 2007). The simulator used in their experiments was UCHILSIM, and the application domain was in the development of walking and kicking behaviours in the Sony AIBO robot.

4.5.2.9 Online Adaptation Approach (Floreano and Urzelai 2001)

Floreano and Urzelai present an interesting alternative approach to many of those discussed here. Instead of striving to create a near-perfect translation of behaviours evolved in simulation onto the real robot in its actual environment, they suggest instead evolving for adaptability. Once an initial neural controller has been evolved, it then quickly learns to adapt on-line to any specific environmental properties, or to any imperfections in the simulator. They suggest also that this phylogenetic–epigenetic approach will allow the robot to adapt on a continuous basis to unpredictable changing environmental conditions, as might be encountered by a robot involved in planetary exploration. These evolved plastic controllers were applied to a Khephera robot in a sequential "light-switching" task.

4.5.2.10 Minimal Simulations and the Envelope of Noise Approach (Jakobi 1997a, b)

Jakobi's radical envelope-of-noise approach involves the separation of the aspects modeled by the simulator into two different types: the base-set of interactions between the robot and the environment corresponding to the core aspects of the simulation which we hope to be able to model with high accuracy, and the implementation aspects of the simulation which cover other areas of the simulation on which the evolving controllers may come to depend. The base-set of robot–environment interactions must be identified and each of the base-set aspects must be varied with each trial to ensure robustness, as it will not be possible to model these interactions with total accuracy.

Sufficient noise must also be added to the implementation aspects of the simulation so that controllers evolve that are not dependant on these aspects for their survival in the evolutionary process. Jakobi applied his approach to two different problems involving a Khephera robot in a T-maze environment and for a visual discrimination task; in both cases the evolved robots successfully traversed the reality gap with highly robust behaviour displayed on the real robots. Issues that arise with this approach include the optimal selection of the base-set aspects and the selection of the level of variation/noise to be applied between experiments.

4.5.2.11 Sensor and Actuator Tuning with the Addition of Noise, and Additional Evolution on the Real Robot (Miglino et al. 1996)

This classic approach involves the tuning up of sensors and actuators as used by the simulator to more accurately reflect the real actuators and sensors on the robot, the addition of small amounts of noise to encourage robust behaviours, and allowing for the continuation of the evolutionary process on the real robot, if deemed necessary.

In their own words:

> We will show that: (a) an accurate model of a particular robot-environment dynamics can be built by sampling the real world through the sensors and the actuators of the robot; (b) the performance gap between the obtained behaviours in simulated and real environment may be significantly reduced by introducing a "conservative" form of noise; (c) if a decrease in performance is observed when the system is transferred in the real environment, successful and robust results can be obtained by continuing the evolutionary process in the real environment for few generations.

The experiments were conducted using a Khephera robot, and the application domain was obstacle avoidance combined with high speed movement.

4.5.2.12 Scaled Experimentation

Experimentation in a simulator also involves a reality of sorts (if simulation is not just a "mind-game"), where the physical movements of the robot are represented in the form of electrons moving, or of the transit of disk read–write heads.

Considering this, we suggest that one approach to partly bridging the reality gap for engineers interested in applying evolutionary techniques to the design of adult-sized humanoid robots might be to build a scaled-down faithful replica of this robot on which to perform these experiments. This scaled-down model would be less likely to cause damage to its environment and would be considerably cheaper to maintain or, in extremis, to replace. Of course, there are aspects of the dynamics of the robots movements which will not scale; however, this approach might provide a useful bridging device when accompanied by judicious experimentation in simulation.

The scaled-down humanoid need not, of course, incorporate the full function-ality of the adult-sized humanoid in terms of sensory apparatus, computational resources, etc., but just needs to incorporate such functions as are required for the experiments in hand. The recent rapid rise in 3D printing technology could prove invaluable in the creation of these small-scale models if such an approach was deemed effective. This approach might also prove a useful "half-way" house between experiments conducted entirely in simulation, and those conducted on a real adult-sized humanoid.

A specially constructed scaled environment could also be constructed for the miniature robot to operate in, in order to demonstrate its domestic (or other) skills.

4.6 Conclusion

We now conclude this short survey of humanoid robots, their simulators, and their connection via the dreaded reality gap. It is probably fair to say that until a situation arises where it becomes possible to automatically construct near-perfect simulations with minimal cost, or it becomes feasible to follow Turing's vision of humanoid robots roaming at will and learning as they go, the reality gap issue is here to stay, even if over time it becomes more of a chink than a chasm.

Of course, not all of the approaches discussed here may be of equal utility in the design of humanoid robots. As pointed out in Koos et al. (2013) in relation to the application of Jakobi's minimal simulation approach to the design of an 8-DOF quadruped,

> Jakobi's methodology can hardly be envisaged for such an application, as it is difficult to define a set of relevant parameters in simulation whose variations would lead to robust controllers.

It is to be expected that a similar constraint would apply in the application of this technique as we apply evolutionary principles to the design of a complex many-DOF humanoid robot

Some of the other approaches we suggested may also be difficult to implement in practice. For example, we noted that the grounded simulation learning approach, while interesting, involves directly modifying the simulator, which may prove difficult, especially if dealing with proprietary software, such as Webots. In fact, the authors managed to circumvent this issue in their research by instead providing the simulator with modified inputs based on experimentation on the real robot.

We note that the notion of the future near-automatic generation of faithful models of real-world artifacts may not be quite as farfetched as it seems at present. Astonishing progress has been made over the last few years in the use of inexpensive hardware devices such as the Kinect sensor for modeling a wide range of real-world environments. Additionally, it should be noted that with the recent rapid advances in 3D printing and associated technologies much progress has also been made in the other direction, that is, the automatic construction of real-world robots and other objects from simulation.

The gap between model and reality is, indeed, beginning to close.

Chapter 5
Evolutionary Humanoid Robotics (EHR)

5.1 Background to, and Motivation for, Evolutionary Humanoid Robotics

Evolutionary humanoid robotics (EHR) involves using a process of artificial evolution to develop some or all of the body and/or brain of a humanoid robot. By definition, obviously, the gross anatomical structure of the robot is predetermined, if we are to take a fairly rigid definition of humanoid, as defined in the Oxford dictionary as

a being resembling a human in its shape

or, when used as an adjective,

having an appearance or character resembling that of a human

However, the way remains open to evolve subtle, yet possibly highly influential anatomical features. For example, just as all humans (except perhaps identical twins) differ slightly or to a larger extent in bodily structure from each other, e.g., leg length, overall height, etc., these particular physical attributes may confer advantages in some areas of physical endeavour, and possible disadvantages in others. For example, longer legs may make one a better runner, however, they may result in a certain awkwardness in confined areas (as the author found to his cost on a crowded Tokyo commuter train). If we are to relax our definition of humanoid slightly, we can include far larger variations in gross morphology and in generating the robot's sensor and motor apparatus (three eyes, four arms, etc.). See Sect. 4.2, "Levels of Anthropomorphicity", for a further discussion of this topic. So, while our main emphasis may be on the evolution of the robot brain, that is the control structures and the memory elements which allow the robot to operate effectively in its chosen environment, we can also place some emphasis on the morphology and

© The Author(s) 2015
M. Eaton, *Evolutionary Humanoid Robotics*,
SpringerBriefs in Intelligent Systems, DOI 10.1007/978-3-662-44599-0_5

sensor and motor arrangements of the humanoid robot. For example, while it may have two "eyes", what exactly is their location, how far are they apart, etc.

Of course locomotion, bipedal or otherwise, is only one aspect of being able to operate effectively in built-for-human (BFH) environments. In order to function effectively in terms of manipulating objects, opening doors, etc., the robot must also have at its disposal a facility of some sort for grasping and manipulating objects.

EHR can be seen as the ultimate goal of the evolutionary robotics movement in some sense. The wheel turns full circle: man, who can be seen, at least in part, as a product of a natural evolutionary process, now produces a humanoid robot/replicant in his/her own image by a process of artificial evolution.

5.1.1 Why Treat Evolutionary Humanoid Robotics as a Special Case of Evolutionary Robotics?

We can appreciate the interest/advantages in evolving humanoid robots in general (because, as we have noted, these robots can operate in BFH environments), but, in particular, what is it that distinguishes evolving morphologies and behaviours for human-like robots in particular (EHR), as opposed to the "general" evolutionary robotics field?

1. As we have previously stated, humanoid robots occupy a special place in so much as, in general, they can operate in environments in which humans would normally be comfortable, that is, built-for-human environments. Additionally they may be able to operate in exceptionally dangerous or challenging environments such as have been created in the aftermath of the Great East Japan earthquake.
2. It has been argued recently that the nature of intelligence has a very close connection with the embodiment of this intelligence. We who wish to build and use these robots (for whatever reason) are humans ourselves. Pfeifer et al. (2007) argue that the nature of the intelligence which is developed is intimately related to the morphology of the robot: "the body shapes the way that we think". The argument is that as humanoid robots are built in a fashion broadly similar to humans, not only does the morphology ("body") of a humanoid robot (more or less) closely resemble that of a human, thus allowing the robot to effectively operate in environments normally populated by humans, but also as a result of this the evolved "intelligent" behaviour should also more closely resemble that of human intelligence.
3. It can also be argued that close morphological resemblance may confer many advantages in terms of human–robot interaction (HRI) in the future, because if robots are broadly humanoid in form, humans may find day-to-day interactions with these robots easier and more natural. However, in this context we must not forget the so-called "uncanny valley" effect (Mori 1970). This effect occurs when the appearance and behaviour of the humanoid robot is sufficiently close

to that of a human that the observer is almost, but not quite, fooled into believing that the robot is, in fact, human. For reasons which we do not have time to elaborate here, this may provoke feelings of dread or apprehension in the human observer, which then vanish as the robot becomes indistinguishable from a human.

5.1.2 Why Are Evolutionary Techniques Not Currently More Widely Applied to Humanoid Robot Design?

A couple of trends become apparent. There has been a huge upsurge in interest in the field of humanoid robotics, especially in the last few years. This author has been astonished at the variety and of the prevalence of recent research. There does not, however, appear to have been a commensurate increase, as one might expect, in the application of evolutionary techniques in the design of these robots.

One significant factor at play here may be the cost. Typical large-scale humanoid robots, for example, the HRP-4C, cost in the region of hundreds of thousands or millions of euros each, and represent many thousands of effort-hours of scientists and engineers working at the forefront of technology in several disciplines. Evolutionary algorithms at present are not regarded by many roboticists as capable of producing demonstrably robust control structures, consequently there is an understandable reluctance to "let loose" an evolved controller on an expensive and difficult-to-repair humanoid robot.

Also, as a consequence of the laws of physics, the larger (and presumably more expensive) the robot, the more serious the potential consequence of a fall or other accident. There is also, of course, the safety of the operators to be taken into account: the larger the robot, the more powerful the actuators, increasing the possibility of a serious mishap. Even the motors of the Bioloid humanoid robot used in our research are capable of delivering a bruising blow if incorrectly programmed— and this robot only stands approximately 40 cm high! We discuss further the issue of the possibility of evolving behaviours initially on a "scaled down" version of a humanoid robot before subsequent transfer to the "full-size" robot, in Chap. 4.

We do, however, see three factors as possibly contributing to the more widespread acceptance of evolutionary techniques in the design of humanoid robots:

- reductions in the cost of underlying hardware as economies of scale begin to kick in
- a greater overall acceptance of evolutionary techniques as they become more prevalent and mainstream in robotics and in other fields
- a growing recognition of the powerful ability of EAs to generate convincingly lifelike behaviours and in generating complex motions which would be impossible to hand-craft easily.

5.2 Approaches to EHR

5.2.1 Evolution of Humanoid "Brain"

By far the majority of the experiments described in this book describe the evolution of the brain of the humanoid robot, by which we generally mean the underlying controller which directs the actions of the robot in response to external stimuli. A commonly used methodology in this regard is to use a genetic algorithm in order to determine the weights and/or topology of a neural network controller in order to effect the desired controller behaviour (Eaton 1993a).

5.2.2 Evolution of Humanoid "Body"/Coevolution of "Body" and "Brain"

It may seem strange that in a text on evolutionary humanoid robotics we include a section on the evolution of the morphology (body) of the robot, as opposed to the robot brain—surely if the robot is of a humanoid form the morphology will be fixed?

This is true on one level, however, as we have briefly discussed previously, just as there are different human body shapes (from child to adult, athletic, muscular, etc.) and different levels of strength versus dexterity and so on, these factors may also have the potential for variation in their robotic counterparts. However, and perhaps more importantly from our perspective, if we take our loose definition of humanoid as any robot designed to operate in an environment designed (primarily) for humans, it becomes evident that a far wider variation in morphological structure becomes possible.

For example, this approach may allow us to include robots with two (or more) alternative methods of locomotion. An example where this facility might prove effective is in an environment which includes stairs, where bipedal or some similar locomotive ability is required, but which also includes flat open spaces of large dimensions as compared to the robot, where rapid wheeled locomotion would prove an advantage. For example, a robot working as an operative in a large hotel complex might well find advantage in having both of these locomotive mechanisms available.

Typically, when evolutionary techniques are applied to the development of aspects of a robot's morphology, they are also employed in the development of the control system. A classic (and much referenced) example of the application of this approach is in Sims' seminal work dating from the early 1990s (Sims 1994a, b).

5.3 Overview of Main Application Domains

5.3.1 Bipedal Locomotion

It is probably fair to say that still the single most common application of evolutionary algorithms to legged humanoid robots is in the evolution of walking or running behaviours. In our tables summarising developments in the field since the year 2008, at least a third of the applications were of this type, and going back further in time the evolution of bipedal locomotion was by far the commonest application.

As Bongard (2013) states in his recent survey of the state of the art in evolutionary robotics

> Many roboticists choose to model the human animal: a humanoid robot is more likely to be able to reach a doorknob, climb steps, or drive a vehicle than a wheeled robot or one measuring only a few inches in length. The humanoid form, however, requires mastery of bipedal locomotion, a notoriously difficult task.

In their recent survey of the field of gait optimisation for bipedal and multipedal legged robots, Gong et al. claim that evolutionary computation is a natural choice for these applications because of the robust and strong global search capabilities of these algorithms together with their ability to deal with situations where a precise model of the robot may be difficult to construct, as may well be the case with a highly complex humanoid robot. They also cite the ability of EAs to deal effectively with multiple objective and multiple constraint problems, and their inherent ease of parallelisation. They claim that because EAs are biologically inspired they may also create more biologically plausible robot behaviours (Gong et al. 2010).

Many factors may be taken into account when designing fitness functions for bipedal locomotion. Two obvious factors are the stability of the robot (i.e., it doesn't fall over, or exhibit unstable walking patterns) and the speed of locomotion. This is often measured by the distance travelled by the robot in a given experimental timeframe; it may also be specified that locomotion in the forward direction only is to be taken into account, to avoid backwards or sideways walking behaviours. Simple measurement of distance travelled in some form falls under the general realm of aggregate fitness functions, as we are just measuring task completion, as opposed to examining particular sensor–actuator behaviours (Nelson et al. 2009). A common criterion for the verification of dynamic stability for a robot in simulation is the zero-moment-point (ZMP). This criterion involves the evaluation of the position where the entire foot needs to be placed in order to have zero moment in the horizontal direction, a criterion which has a long history in humanoid robotics research (Vukobrotovic and Borovac 2004).

Another very common factor used in the creation of fitness functions for bipedal locomotion is the energy consumed by the robot. This has the twin advantages of potentially resulting in more humanlike locomotion, and also in the reduction of power consumption, which is an important consideration to be taken into account, especially if the robot is running on battery power alone (Gong et al. 2010).

Minimisation of torque change and smoothness criteria may also be components of fitness functions. Fitness may, of course, also be determined by a process of interactive evolutionary computation (IEC) in which a human observer(s) either wholly or partly replaces the objective fitness function (Takagi 2001). However, for bipedal locomotion IEC is not so commonly employed, mainly because of the ready availability of objective evaluation criteria, as opposed to dance, for example, because of the highly subjective nature of peoples' evaluation of dance performances. See Figs. 5.1 and 5.2 for examples of evolved walks on a humanoid modeled broadly on the (discontinued) Sony QRIO robot, and on a model of the Aldebaran Robotics Nao humanoid.

5.3.2 Dance

After bipedal locomotion it is probably fair to say that the generation of dance behaviours is currently one of the most common applications of evolutionary algorithms in the design of humanoid robots. Unlike bipedal locomotion, where the fitness function is normally objective by nature, in dance generation a subjective fitness function is commonly employed using a process of interactive evolutionary computation (Virčíková and Sinčák 2011; Dubbin and Stanley 2010). This is mainly because of the because of the difficulty in formulating a fitness function which can extract the preferences of human observers in evaluating dance performance; however, some attempts have been made to formulate an objective fitness function in this regard (Eaton 2013). See Fig. 5.3 for an example of an evolved sequence of dance moves using the Webots simulator for the Bioloid humanoid,

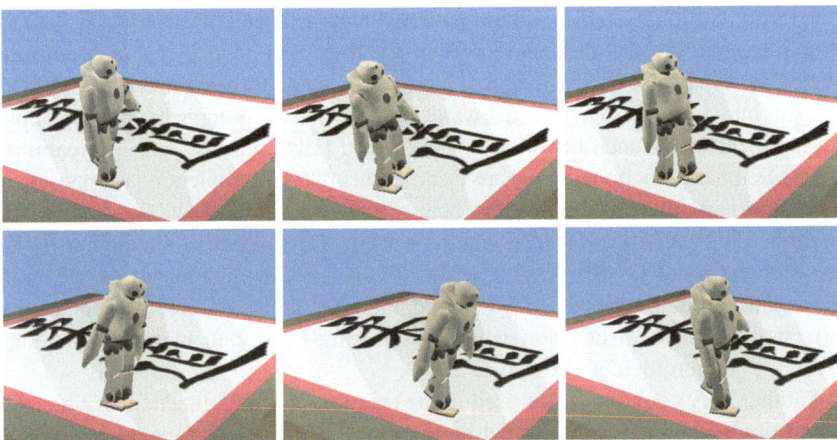

Fig. 5.1 An evolved simulated robot walking with a limping gait based on the Sony QRIO robot, and simulated using the Webots simulator

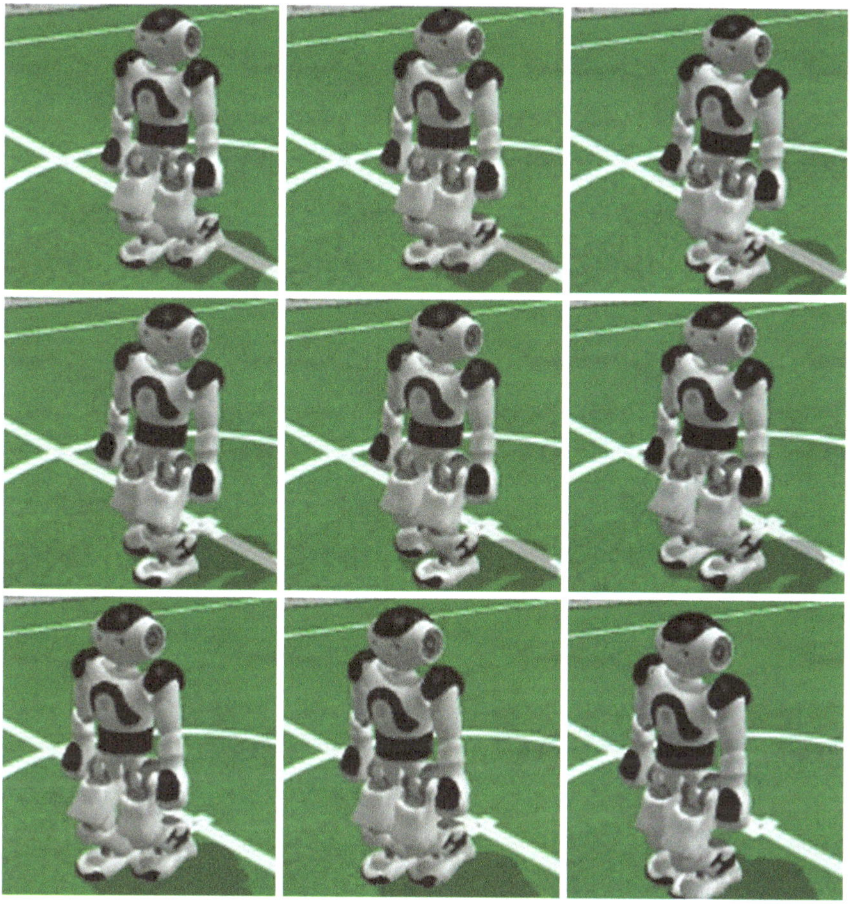

Fig. 5.2 The Alderbaran robotics Nao robot in a simulated walk (Torres and Garrido 2012). Images read from *top left* to *bottom right*

Fig. 5.3 A sequence of evolved dance moves as generated in the Webots simulator using a model of the Bioloid humanoid

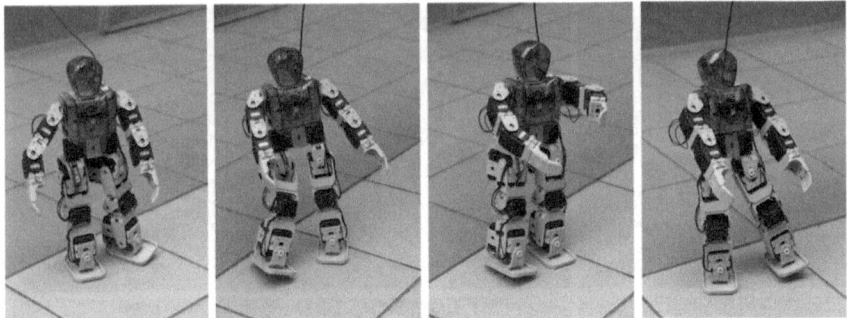

Fig. 5.4 The sequence of evolved dance moves from Fig. 5.3 as transferred to the actual Bioloid humanoid robot

and Fig. 5.4 for this sequence of dance moves, as transferred to the real robot. Of course issues of stability of the evolved robot motions apply to dance as to bipedal locomotion, whether this issue is addressed by a human observer or otherwise.

5.3.3 Other Locomotive Skills—Kicking, Crawling, Jumping, Ladder Climbing

In recent years, since the Aldebaran Robotics Nao humanoid robot replaced the Sony Aibo puppy robot as the robot used in the RoboCup Standard Platform League (SPL), the number of applications of evolutionary algorithms in the generation of soccer playing skills such as kicking, walking, and turning has risen dramatically (Urieli et al. 2011; Gokce and Akin 2011). We expect this trend to continue into the future, as the emphasis will move from the evolution of bipedal locomotion to more complex multifunctional behaviours. Figure 5.5 shows an example of an evolved jump from a reduced gravity run (simulating moon-like gravity) using a simulated QRIO-like robot, and Fig. 5.6 gives an example of evolved ladder climbing using a simulated humanoid.

5.3.4 Grasping and Manipulation

While the area of grasping and manipulation is still one of the less-common applications of evolutionary robotics in the design of humanoid robots, it is of crucial importance in the design of future humanoid robots designed to interact with their environment in an intelligent and useful fashion. An interesting recent application of evolutionary robotics in this field was that by Savastano and Nolfi, who used an incremental approach to the development of reaching and grasping behaviours in the iCub humanoid robot. This work could also fall under the

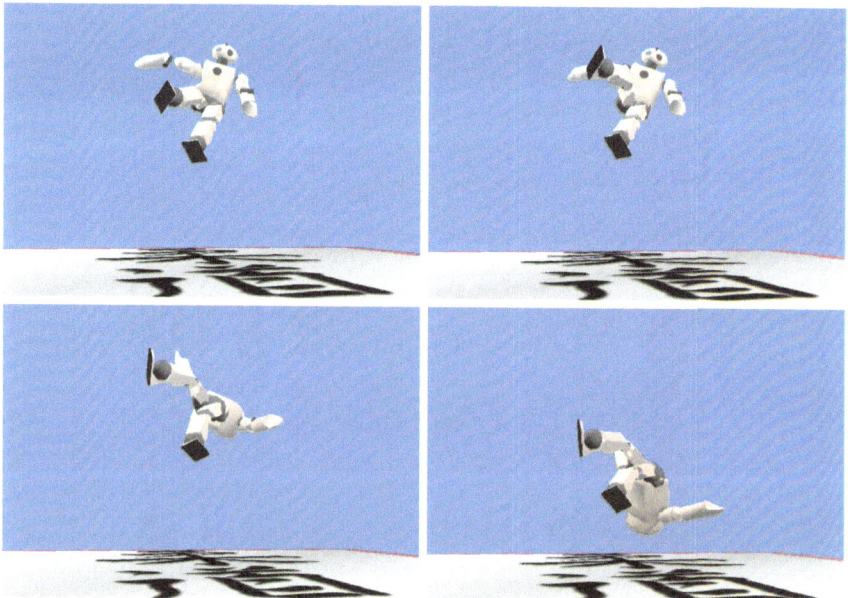

Fig. 5.5   An example of an evolved jump from a reduced gravity run using the simulated QRIO model

category of the investigation of human motor skills as the experiments conducted were designed to mimic the experimental settings in which infants are studied. Performance level, in a basic sense, was measured as the minimum of the distance between the centre of the robot's visual field and the number of fingers in contact with the object, both values scaled to between 0 and 1. The researchers noted that the skills evolved by the simulated iCub robot were similar to those displayed by real children (Savastano and Nolfi 2012). Figure 5.7 demonstrates an early application of the evolution of grasping behaviour.

5.3.5   The Investigation of Human Motor/Locomotive Skills

Another interesting application of evolutionary techniques applied to simulated or embodied entities of human likeness is in the investigation of human motor skills. For example, Sellers and colleagues conducted an investigation into the importance of the Achilles tendon in human running ability. They suggested that identifying which early hominid species have this tendon would give a good indication of their locomotive abilities (Sellers et al. 2010; Fig. 5.8). In other work Sellers investigated using a genetic algorithm in order to evaluate alternative gait strategies for early hominids (Sellers et al. 2004), and in predicting the most energy-efficient upright

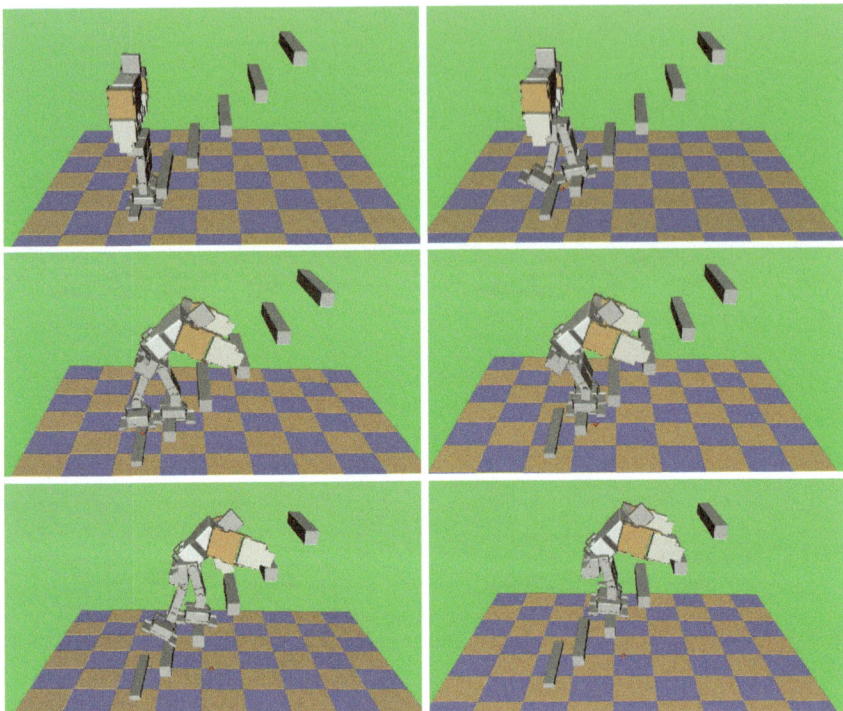

Fig. 5.6 An example of evolved ladder climbing in a simulated humanoid robot (Wang et al. 2012). Images read from *top left* to *bottom right*

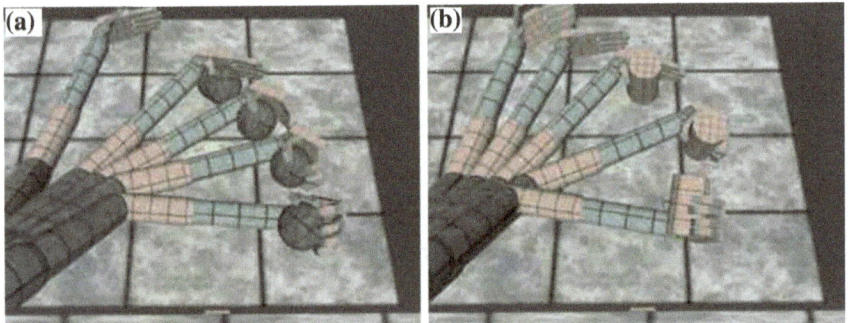

Fig. 5.7 An example of the evolution of grasping behaviour in simulation (Massera et al. 2007). These images show five superimposed snapshots of the evolved behaviour (using the same evolved robot), for **a** grasping a sphere, and **b** grasping a cylinder

walking gait for *Australopithecus afarensis*, an early human relative (Sellers et al. 2005). They also used evolutionary techniques in the investigation of the maximum running speeds of bipedal species of dinosaurs (Sellers and Manning 2007).

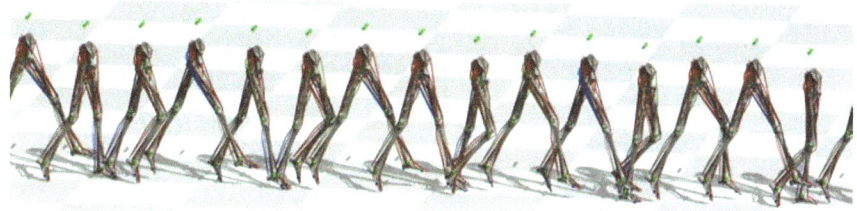

Fig. 5.8 A composite picture of an example evolved simulated gait, sampled at 0.1-s intervals, using a high-quality musculoskeletal model (Sellers et al. 2010). The checkerboard spacing is 1 m by 1 m

Another early piece of work in this regard is Hase and Yamazaki's investigation into the evolution of walking behaviours in nature. An interesting aspect of their work, which is discussed in more detail in our "prehistory" section, is that Hase and Yamazaki (1999) demonstrate that, starting from walking patterns closely resembling those of chimpanzees, as evolution progressed there was a shift in both morphology and control from apelike walking to more humanlike walking. Hase et al. further discussed this topic in 2003, and they suggested that their work might form the basis for rehabilitation tool design for persons suffering from physical handicaps (Hase et al. 2003).

5.3.6 Other Application Areas

Some of the more unusual recent applications include the evolution of yoga positions (Hettiarachichi and Iba 2010, 2012; Figs. 5.9, 5.10 and 5.11), ball catching (Ra et al. 2008), and performing as an exhibition guide robot (Fukunga et al. 2012;

Fig. 5.9 Three different target motion trajectories for yoga (Hettiarachchi et al. 2012). On the *left* is the warrior pose, the *middle* pose is called the frontal leg raise, and the one on the *right* is the hand-to toe pose

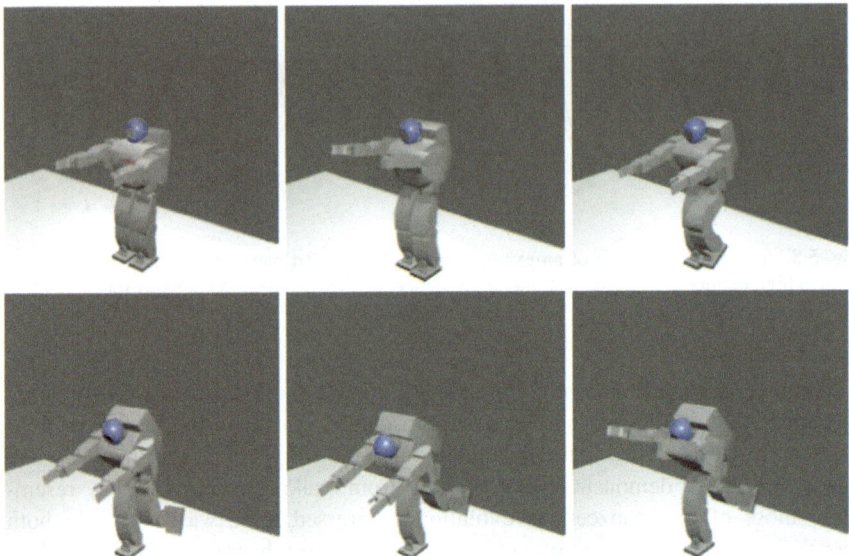

Fig. 5.10 The behaviour of the best evolved individuals in the simulator environment for the warrior pose (Hettiarachchi et al. 2012)

Fig. 5.12). We include this final example as, although wheeled, the robot in this study is designed to operate in built-for-human environments, albeit in a limited fashion, and is one of the few examples of the application of evolutionary techniques in the evolution of a commercial adult-sized humanoid robot operating in a real-world environment. A prototype of this robot in the real world successfully guided students posing as visitors around an exhibition space with four poster "exhibits".

5.4 Commercial Applications of EHR

Two figures stand out in the application of EHR techniques, albeit in simulation, to commercial applications: Karl Sims (although admittedly Sims' work was not particularly targeted at humanoids), and Torsten Reil, both of whom went on to found successful companies. Karl Sims founded GenArts based in Cambridge, Massachusetts, a company involved in the generation of special effects for films. Torsten Reil went on to found NaturalMotion in 2001, a company based in Oxford and San Francisco, which specialises in animation software engines for games and films. In a recent development, NaturalMotion was acquired (in January 2014) by Zygna, a social network gaming company, for a figure estimated at over half a

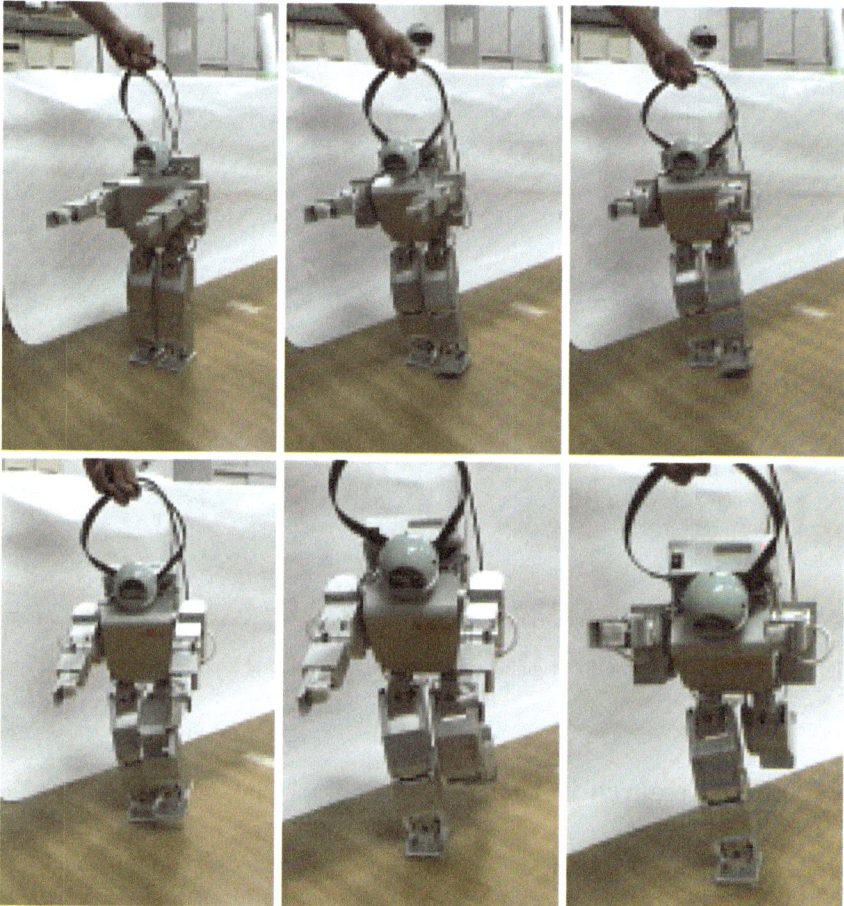

Fig. 5.11 The warrior pose as shown in simulation in Fig. 5.10 as applied to the real HOAP-2 humanoid robot (Hettiarachchi et al. 2012)

billion US dollars. Another early pioneering researcher in the field, Peter Nordin, was involved in the creation of several successful companies, and has been listed as one of Sweden's 12 most influential inventors by the Swedish Trade Council. As the potential of evolutionary algorithms for the generation of realistic human motions becomes more fully realised we expect a commensurate increase in commercial applications, both in simulation (for the games and film industries) and in their application to real humanoid robots.

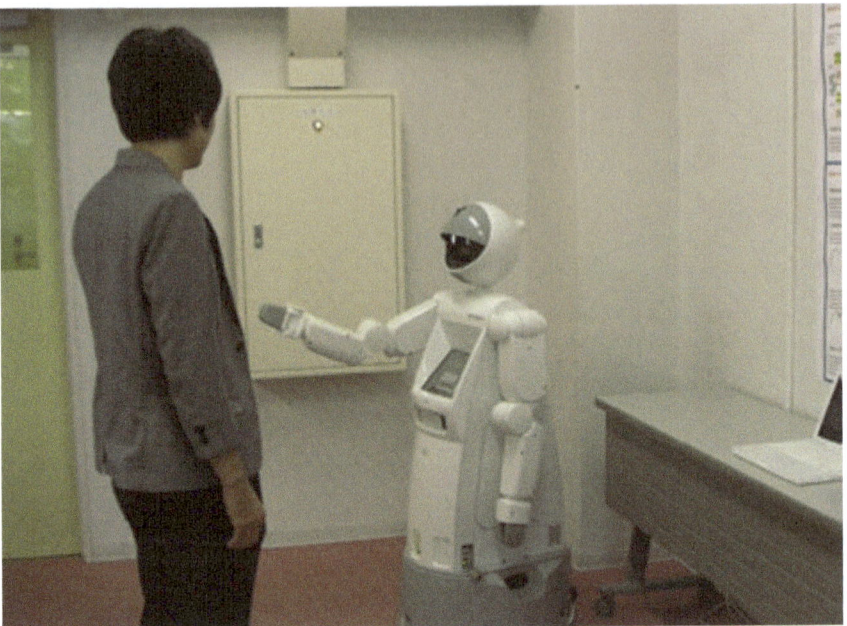

Fig. 5.12 The Fujitsu enon robot guide acting as a visitor guide at a poster exhibit (Fukunaga et al. 2012)

5.5 Initial Research in EHR

In this section we discuss some of the many early attempts, some fruitful (and some perhaps not so much so), by researchers seeking to apply ideas from the principles of natural evolution to the design of both body and brain of humanoid robots. Because of the sheer volume of research conducted in the area of evolutionary robotics since the early 1990s it is necessary to draw certain boundaries around the research developments described here. For example there has been a considerable body of work done concerning the evolution of multipedal robots of the nonbipedal type. Indeed, the experiments performed by Lewis et al. at the University of Southern California on the interactive evolution of locomotion for a hexapod robot in 1992 are considered by Nolfi and Floreano in their book *Evolutionary Robotics* to be probably the first experiments conducted in evolutionary robotics. Although this body of work obviously has close connections with the evolution of bipedal locomotion, it was considered necessary to omit reference to all but the most important and influential of these works for space reasons.

Also, a large body of work has been done on the evolution of humanlike behaviour in simulation only—some of this work has close connections with the artificial life (Alife) field. It was considered important for completeness to include reference to some of this work (much of this work was conducted in a real-physics simulation environment, so there is, indeed, the possibility of future transference to real humanoid robots, while acknowledging the presence of the reality gap issue). However, not as much weight has been given to this work as to that research in which the evolutionary process ends up on a real humanoid robot, either through an evolutionary process taking place in a simulator for transference to the real humanoid or (much less commonly) the evolutionary process taking place completely on the real robot. It is also possible to follow some "in between" process, perhaps involving initial evolution taking place in the simulator before transference to the real robot for a "finishing off" process. Indeed, much of the work involving simulation only, while producing very lifelike and realistic motions, is never intended for transference to a real robot, but is for use in the computer graphics industry for applications in computer games and in film production.

We also discuss work involving the evolution of both the morphology and the control circuitry of a robot or simulated creature; while the morphology of the evolved creatures may not initially be humanlike, it may have the potential to evolve in this direction. Sims' early work on the evolution of virtual creatures comes to mind in this regard, together with the important work over many years by Hod Lipson, Josh Bongard, Jordan Pollack, Bob Full, and many other researchers. In fact, this line of research can be said in a sense to have the most potential, as not only are we evolving the control structures of the robot, we can also alter the morphology in either a gross or a minute fashion. For example, we can change the robot's height or weight, lengthen its arm a little to allow for greater reach, or, on the other end of the spectrum, add an arm here, a leg there, or an extra set of vision receptors on some other part of the robot's body.

5.5.1 EHR "Prehistory"—Developments in EHR Prior to 2000

5.5.1.1 Introduction

In this section we look at some of the earliest attempts to apply evolutionary techniques to robots which we may describe as humanoid in some sense. Of course the availability of sophisticated computer hardware was limited and the state of the art as regards robotic technology was far behind what we have today. ASIMO first appeared in the year 2000, although it did have many precursors back to the E0 in 1986; indeed Waseda University's WABOT-1 first appeared in 1973. However, there was some remarkable and seminal work done in this period which, it is fair to say, still influences current work in this field. One piece of research which stands out to this day was Sims' 1994 research on the evolution of artificial creatures. Although this work probably belongs more in the Alife field than in EHR per se, it has had a remarkable influence, in that it involved the evolution (albeit in simulation) of artificial creatures of fully three dimensional form which could perform a variety of functions including swimming, walking, jumping, and following (Sims 1994a). Both creature morphology and control circuitry were evolved, and although the creatures evolved were not specifically humanoid in shape, it should have been possible to bias the evolution in this direction if this was so desired. In a separate piece of work Sims investigated the evolution of 3D morphology and behaviour by competition. It should be noted that all of this research was fully noninteractive in nature; that is, no human intervention was required in the selection of individuals for the genetic process. This is in contrast to Dawkins' (1986) early work on Biomorphs, described in his book *The Blind Watchmaker*, in which the selection of individuals was based on the user's aesthetic preference rather than automatically selected by a predefined fitness function.

Probably the earliest descriptions of the application of evolutionary techniques to the evolution of bipedal locomotion were de Garis's (1990a–c) articles on the application of what he called "genetic programming", which he used to send control signals to a pair of stick legs to teach them how to walk. Other research of note in these early "prehistoric" days includes Gritz and Hahn's (1995, 1997) early work on genetic programming for articulated figure motion, and Arakawa and Fukuda's work on the application of a genetic algorithm for motion trajectory generation for a biped robot. This was probably one of the earliest applications of evolutionary algorithms to motion control of a real humanoid robot (Arakawa and Fukuda 1996, 1997).

5.5.1.2 The Experiments

One of the earliest researchers to consider the application of evolutionary techniques to the generation of bipedal locomotion is Hugo de Garis (1990a–c). He used a technique which he termed genetic programming or

> the application of the Genetic Algorithm to the evolution of the signs and weights of fully (self) connected neural network modules which perform some time (in)dependent function (e.g., walking, oscillating, etc.) in an "optimal" manner

This technique was used in order to evolve neural network modules (called GenNets), which made a pair of stick legs move as far as possible in a specified number of cycles. The length of the cycle time was also evolved by the GA. The idea was that once the signs and the weights of individual modules were determined these would then be "frozen", and this module could then be used as a component in a more complex structure. Eight input neurons were used, taking as input the values of the angles and of the angular velocities of the "hip" and of the "knee" joints of each of the two stick legs; the four output neurons provided the angular acceleration of each of these four joints. Weights were encoded as binary strings, using standard roulette wheel selection and only the mutation operator was allowed, with no crossover. De Garis argued that because of the high interdependency of the neurons, the crossover operator would have a negative effect on the evolutionary process.

Initial experiments without imposing any constraints on the movements of the legs produced movement, however, this was highly unlifelike and involved windmilling motions and strange contortions of the joints. However, when the experiment was performed on the simulated stick legs using a full set of constraints to try to replicate realistic walking motion the evolutionary process failed, producing stick legs which took one giant step, did the "splits", and then ceased to move. In order to counter this, de Garis proposed the idea of "sequential evolution" in which separate phases of evolution would be used, each with its own individual "quality measure" or fitness function; each stage of the evolutionary process taking as its starting population the evolved behaviour from the previous phase. This was based on his observation that behaviours evolved in the early stage of an evolutionary process tended to persist into later phases of evolution, a concept he termed "behavioural memory".

Three phases were used in all; the first two phases involved tailored fitness functions, and the final phase used just the distance covered as the fitness function. At the end of this sequential evolutionary process the stick legs were walking. De Garis argued quite strongly in these articles that in the future as genetically evolved modules increased in complexity, it would be necessary to abandon any hope of having an understanding of the internal operation of these modules, and that in the future only their level of performance at the task in hand would (or should) be of concern.

De Garis (1990a) also suggested here the possibility of implementing the evolutionary process on real robots

> One can imagine a succession of experiments being performed on a single robot or a single experiment on a family of robots

and on the precautions that might need to be made in this case,

> Failsafe precautions could be implemented into the robot while it evolved, so that it would not be damaged by falling to the ground

and concludes this article on the optimistic note,

> the GA may provide solutions to problems which remain unsolvable with other techniques. After all, it is the preferred approach used by nature against its gargantuan complexity problem.

In 1992, inspired by de Garis's work and by earlier work by Beer et al. (1990) which demonstrated that a network of simple neurons could produce locomotive behaviour in a simulated artificial insect, Lewis et al. (1992) at the University of Southern California conducted research in the evolution of control signals to drive a physical 12-DOF 6-legged insect robot. Similar to de Garis's work, Beer et al. used a process of incremental evolution which they called "staged evolution", a process which could be today referred to as employing a "functional incremental fitness function" as defined in Nelson et al.'s comprehensive 2009 taxonomy of fitness functions (Nelson et al. 2009). This process also involved the experimenter explicitly scoring behaviours generated by the robot in a process of interactive evolutionary computation, so we may refer to the evolutionary process as employing a functional incremental interactive fitness function, or just an incremental interactive fitness function for short.

The first stage of the evolutionary process involved creating a neural oscillator to drive a leg circuit, and the performance of this oscillator was evaluated by a qualitative visual inspection by the experimenter, who is looking to see a consistent oscillatory behaviour. The evolved oscillatory circuits were rated on a scale between 0 and 60, and once at least half of the population scored near 60 the second stage of evolution commenced. For the second stage, the set of connections between the oscillators was also specified and was subject to the evolutionary process. The second stage of evolution also involved visual evaluation by the experimenter, however, this time based on more objective measures, including the distance walked by the robot in the forward direction, and the number of degrees turned by the robot during its walk. In all of the experiments performed, tripod walking behaviour evolved; this is a gait seen on many insects. Unlike de Garis's experiments, crossover was also employed with a probability of 0.01, and the probability of mutation was set at 0.04. The experiments used a 65-bit string to encode the different parameters (eight bits for each of eight parameters, plus another bit to specify forward or backwards walking) and used the Gray coding scheme.

In another 1992 piece of research, Beer and Gallagher (1992) also evolved locomotive behaviour for a six-legged insect, however, in simulation only. In their case a 200-bit string was used to encode the parameters of a recurrent neural network with 50 free parameters. Beer and Gallagher tried the incremental evolutionary approach espoused by de Garis in his earlier (1990a–c) work and in the fashion used by Lewis et al. in the same year, but found that the performance of controllers evolved in this way in general displayed inferior performance to those evolved "from scratch". They used a noninteractive approach to the evolutionary process, and the fitness function used was of the aggregate type, based only on the forward distance moved by the insect in a fixed amount of time. A variety of experiments were performed; in some of these experiments sensory input was provided in the form of leg angle sensors, while in other experiments the robot was deprived of this feedback. In all of the evolutionary runs performed, a tripod gait was evolved which was observed to pass through four distinct evolutionary phases: initially just grounding all six feet and pushing until they fell over, then passing through a phase of swinging their feet in an uncoordinated fashion until they fell over, albeit having made some progress. The final two phases involved stable but uncoordinated gaits and finally stable coordinated locomotive patterns.

Karl Sims was one of the first researchers to coevolve both body and brain of three dimensional simulated "creatures" in his now near-legendary and highly influential 1994 article "Evolving Virtual Creatures" (Sims 1994a). Although the goal of this research was not to evolve and build real robots, it has inspired other research aimed at producing real robots with variable morphologies. It should be noted that while no attempt was made to evolve specifically humanoid like creatures, this would not have been beyond the bounds of possibility. Each individual creature is represented by a directed graph of nodes and their connections.

The structure of the creature's body is assumed to be composed of articulated three-dimensional rigid components (simple rectangular solids) which are allowed to overlap at the joints. Each node in the graph contains information about the dimensions of a rigid component, the type of joint which connects this part to its parent node in terms of the number of degrees of freedom, and the type of joint movement together with individual joint limits. Each node also contains a set of local neurons and information about its connectivity to other nodes. The set of computations available to each individual is varied and ranges in complexity from simple addition to functions generating more complex oscillatory outputs. The number of inputs to each neuron varies from one to three. The complete set of allowed functions is as follows (Sims 1994a):

> sum, product, divide, sum-threshold, greater-than, sign-of, min, max, abs, if, interpolate, sin, cos, atan, log, expt, sigmoid, integrate, differentiate, smooth, memory, oscillate-wave, and oscillate-saw.

Sims allowed for three basic types of sensors: joint-angle sensors which detected the current angle at each joint for each of its degrees of freedom, contact sensors for collision detection, and photo sensors for light detection in the simulated environment. Motor effectors generated forces on joints in order to move the creatures around in the various types of locomotion which they evolved. Walking behaviour (a term Sims used to describe any form of locomotion on land) was evolved using a simple fitness function based on measuring the distance travelled by the creature's centre of mass over a period of time. Jumping behaviour was evaluated using two alternative fitness functions; one involved the maximum height above the ground plane reached by the lowest part of the creature, while the other method involved using the average height of the lowest part of the creature above the ground over the simulation period as the fitness measure. To simulate underwater locomotion (swimming) a viscosity effect was implemented in the simulation; while not entirely physically accurate, this was sufficient to emulate swimming and paddling-type behaviours. For swimming the effect of gravity on the environment and the fitness function, similar to walking, was based on the total distance travelled over a period of time by the centre of mass of the creature, also taking into account that the creature can move in the vertical direction.

It should be noted that although Sims' work was strictly in simulation, he did acknowledge that his approach might have application in the evolution of real robots (one of the "creature morphologies" pictured in his paper had a broadly humanoid shape), if constraints were placed on the evolutionary process, so as to only evolve creatures which could be implemented as real robots. Sims also briefly discussed the possibility of using "aesthetic selection" (interactive evolutionary computation) in place of the objective fitness functions he espoused, however, he dismissed this as requiring "too much patience on the part of the user". He did, however, suggest a methodology of interleaving interactive and objective fitness evaluations as a way of reducing this onerous burden on the experimenter. In a separate paper in the same year Sims also described the application of a similar evolutionary process where the standard fitness functions used were replaced by a competitive fitness function where two simulated creatures fought for control of a cube (Sims 1994b). The winner was the one with the shorter distance to the cube after a set period of simulated time. Creatures were thus encouraged not only to approach the cube, but also to keep their opponent away from it. Based on his experiments Sims concluded that

> it might be easier to evolve virtual entities exhibiting intelligent behaviour than it would be for humans to design and build them.

However, he readily acknowledged, as did de Garis, that in the future, complex evolved creatures might not be readily amenable to human understanding

> As computers become more powerful, the creation of virtual actors, whether animal, human, or completely unearthly, may be limited mainly by our ability to design them, rather than our ability to satisfy their computational requirements. A control system that someday actually generates "intelligent" behaviour might tend to be a complex mess beyond our understanding.

In one of the earliest applications of genetic programming (in the more conventional sense of evolving programs rather than as described by de Garis in 1990) to the control of a simulated humanoid, Gritz and Kahn (1995) applied this technique to generating control programs for simulated robot controllers. Rather than evolve both controller and morphology, as in Sims' earlier work, their work solely concentrated on the evolution of controllers for fixed topology structures. Their aim was to attempt to automate the animation process for the generation of articulated figures for character animation applications. The fitness functions employed involved two separate components—a main goal and a set of so-called style points. The initial application was in the articulation of a model of a lamp with three controllable degrees of freedom. The function set for the genetic programming process consisted of just the operators {+, −, *, %, ifltz}. Here +, −, and * are the three standard arithmetic operators of addition, subtraction and multiplication; % is the protected division operator—the same as standard division, however % does not cause an exception when dividing by zero. Finally, ifltz evaluates a given expression, and if this evaluates to less than zero, one action is taken, otherwise a different action is taken. For this experiment the objective was to move the lamp from its initial position to a particular spot on the floor, and the main goal component of the fitness function involved the distance between the centre of the lamp and the goal point at the end of a set time period. Style points included a bonus for early completion of the task, and penalties for falling over, or for excess movement after reaching the goal. An incremental learning approach was used, with just the main goal used as the fitness function early in the evolutionary process, and the style points were gradually incorporated as the evolutionary process progressed. The final controller evolved using this process resulted in a hopping motion for the lamp, bringing it to the precise spot required. These first experiments just involved motion to a particular spot, but later experiments aimed at producing a generalised locomotion controller, which could move the lamp to any desired position (Gritz and Kahn 1997).

This approach was then extended to the control of a simulated humanoid figure with a total of 28 degrees of freedom, where between 4 and 10 of these DOF could be under genetic control at any one time. The terminal set included the position of the humanoid relative to the goal point, the time, the value of internal joint angles and force sensors and the positions of end effectors, together with a set of randomly chosen floating-point constants. Using the same function set as in the previous experiment, and fitness functions constructed in a similar fashion (a main goal in conjunction with a set of style points) a variety of nontrivial humanoid actions were evolved, including pointing at objects and touching them, making gestures, and touching its nose and other parts of the body. The evolved motions were considered to be fluid and lifelike, and entertaining to watch, an important consideration for character animation. Gritz and Kahn did, however, acknowledge the brittleness of the evolved controllers, and the difficulty in the generation of good fitness functions, also acknowledging that the first fitness functions were generated through a process of trial and error. This is an issue discussed further in Sect. 3.3.2 of this text.

In one of the earliest experiments demonstrating the application of evolutionary algorithms to a real humanoid, Arakawa and Fukuda (1996) describe the use of a genetic algorithm to determine joint angles for bipedal locomotion for a real bipedal robot with 13 joints, constructed from aluminium. The fitness function employed was based on minimising energy consumption subject to a number of imposed constraints, to avoid the robot falling over. Following evolution in simulation, the evolved trajectories were successfully transferred to the real robot resulting in a walking speed of 0.3 m every 5 s. Further evolutionary experiments using this physical robot platform are described in Arakawa and Fukuda (1997), Hasegawa et al. (2000).

Another early work on the application of evolutionary algorithms for biped locomotion was Cheng and Lin's (1995) application to a simulated five-link biped model. The joint angles for four of the biped's links were specified by the genetic algorithm; the fifth joint angle representing the deviation from the upright of the simulated biped was set to zero in order to promote upright motion. As with much research around this period using simulated bipeds, only motion in the sagittal plane is considered. The positions in time for the robot's joints for six instants in time are specified, and transitions between these points are smoothed by quadratic polynomial functions. Several experiments were conducted using different fitness functions to select between different locomotive behaviours, including total walking time without falling (up to a maximum of 100 s), duration of walking combined with average body speed (to promote high-speed walking), and selecting for a specific step size. Walking was also evolved for both an inclined and a declined sloping surface. An extension of this paper in 1997 tested for an additional two sloping surfaces (Cheng and Lin 1997).

In 1998 Juárez-Guerro and colleagues experimented with the generation of a walking gait using evolutionary strategies for a curious-looking biped with a passive "tail" (Juárez-Guerro et al. 1998). Aspects of the morphology of this robot were also evolved, making this one of the earliest experiments to evolve aspects of both the body and the brain of a physically realised broadly bipedal robot (although calling it humanoid would be a bit of a stretch).

Other early work includes using a genetic algorithm to smooth the transitions between user generated "via points" by decreasing the peak values of velocity and acceleration between these points (Choi et al. 1999), and using a genetic algorithm to generate swinging motions for a bipedal robot (Nagasaka et al. 1997). Both of these experiments involved final successful implementation of the evolved behaviours on a real robot.

Finally, in this section we briefly discuss Hase and Yamasaki's (1999) work on the evolution of walking behaviours. This research is interesting in that, unlike much of the previous research discussed, it used a detailed neuro-musculoskeletal model as developed previously by the researchers, rather than utilising highly simplified biped models. The basic movement patterns for bipedal locomotion for this model are generated by calculating the solutions of 64 nonlinear first-order differential equations describing the behaviour of both the nervous system, which consists of 18 neural oscillators, and the musculoskeletal system. Both the final body morphology and the detailed movement patterns are, however, determined by the evolutionary algorithm. Using evolutionary strategies three separate issues were identified: energy consumption, muscular fatigue, and the load on the skeletal system. The evolutionary process was used to minimise each of these components. An interesting aspect of this research is that, starting from walking patterns known to resemble closely those of the chimpanzee, as evolution progressed both the length of bones and their masses increased, especially in the lower extremities, the torso straightened, and in general there was a shift over the evolutionary process from apelike walking to more humanlike walking. The authors suggest that these evolved changes mirror very closely what is considered to have occurred in the natural human evolutionary process, and in particular that the morphology of the evolved creature bears resemblance to Australopithecines, which it is estimated existed about 3.5 million years ago.

5.5.1.3 Summary of EHR Prehistory

This, then, concludes our short tour of the prehistory of research in the area of evolutionary humanoid robotics, that is, research prior to the year 2000. Many of the foundations for future research were laid in this period. A common characteristic of much of the research in this period is that much of the simulation work was done using numerical calculations without the creation of detailed graphical representations of the behaviour of the robots evolved. This was mainly, of course, due to the limitations in the computing power available at this time. A notable exception is Sims' work on evolving virtual creatures; he did, however, have at his disposal the parallel processing CM-5 Connection Machine, one of the most powerful computers in operation at this time. The videos produced of these evolved virtual creatures represent, to this day, some of the most engaging examples of the power of evolutionary computation to shape both body and brain of simple simulated creatures.

5.6 Simple Illustrative Example: The Evolution of Dance

Here we give a simple example illustrating the possible application of evolutionary algorithms to the evolution of human dance movements. This is based broadly on our recent work on the evolution of dance (Eaton 2013); however, in the case of the research described in this article, the dance was generated in a fully noninteractive fashion. In the example outlined here we include an element of interactive evolution for illustrative purposes.

One of our rationales behind using the evolutionary process to evolve dance movements is to aim for a fluid, rhythmic motion which is pleasing to the eye (this is, of course, to some extent dependant on the eye of the beholder, hence the interactive evolutionary aspect of this work). Generally, in the evolution of robotic dance, no motion, or very little motion would not rate very highly, as would extremely random joint movements. In addition, dances involving the robot dancer falling over midway through the dance would not (generally) meet with high approval ratings! Finally, and this is the more difficult and subjective area, we would aspire to a dance that is in some sense "pleasing" and "artistic".

In order to address the issue of avoiding dances with little or no movement involved we could employ a fitness function designed specifically to reward dance sequences which maximise movement on the dance floor. This could be generated by summing the total movements over a particular dance sequence for each individual joint and then summing these generated values over all of the robot's joints. The approach taken in Eaton (2013) is to actually take the product of each of these individual joint movements, thus promoting at least a little activity in all of the humanoid's joints. This is in keeping with Krasnow and Chatfields' (2009) "performance competence evaluation measure" for the assessment of the qualitative aspects of human dance performance, which suggests that there should be

> No displays of "dead" or unattended body segments when focus of the movement is elsewhere, resulting in all body segments being energized, regardless of how minimal the movement is.

The issue of very random, jerking dances could be addressed in two ways. First, because of the cyclical nature of the movements generated by the genetic algorithm (see Eaton (2013) for a further discussion of this), a level of order is inherent in the robot's movements. Second, very jerky movements may well result in the robot falling over, which results in a sequence being terminated prematurely and the robot being subsequently being assigned a relatively low fitness value.

The third issue, that of the dancer falling over midway through the dance, is also addressed in this fashion (a premature termination of the sequence and the assignment of a low overall fitness to the individual). The final, more subjective, issue of what constitutes a "pleasing" or an "artistic" dance is more difficult to evaluate, and to a large extent is dependent on the human observer.

As, to a certain extent, the perceived quality of a given dance is inherently subjective, we could employ a person or people in the loop. However, in order to reduce the human interaction to a minimum to facilitate an automated process and to reduce human boredom (which might possibly result in "inaccurate" classifications), we could use interactive evolutionary computation (IEC) in combination with the separate objective evaluation of the dance performance outlined above.

This technique involves the majority of the decisions taken in the evolutionary process, i.e., fitness evaluations, etc., being taken by the computer, with the human only being involved at critical stages of the evolutionary process. The human observer is given just a small number of example dances to evaluate, thus reducing the fatigue element for the observer and hopefully resulting in "better" dance evaluations. An example outline of such an evolutionary algorithm is given below.

In this illustrative example the evolutionary process is divided into two distinct stages, involving an objective fitness function (OBJECTIVEEVALUATE), which does not require any human intervention/evaluation, and a subjective fitness function (SUBJECTIVEEVALUATE), involving a process of interactive evolutionary computation (IEC), in which, as we have discussed, the human observer(s) take the part of the objective evaluation function. In the example described here the two stages are invoked in sequence: first the objective evaluation stage which creates a population of humanoid dancers for the subsequent evaluation by the human observer(s) in the IEC phase. Of course these two stages could, instead of following each other in a sequential fashion, be interleaved. In the example here, the user selects the value for the number of generations, this value can also be set interactively, when the user judges that the "objective" phase has run its course. The user also specifies the probability of mutation and of crossover for the evolutionary algorithm. In this example the SELECT operator chooses an individual from the population with a probability proportional to its fitness. A copy is made of this chosen individual, and the CROSSOVER operator then crosses this individual with another individual chosen at random from the current population; the MUTATE operator then applies the mutation operation. This updated individual is then copied into the next generation. This process continues until the next generation is full.

So, starting from an initially randomised population GO(1), the evolutionary process proceeds, and the best ten dances can then be extracted from the final evolved population (this figure cannot be set too high, or else user fatigue may occur in the subsequent IEC phase). These dances then form the initial population for the IEC phase. As for the objective phase, the operators of selection, crossover, and mutation are applied. This phase can continue until the user(s) consider that a dance(s) of sufficient aesthetic quality has been evolved, and the process then terminates with the best evolved dance(s) from the final generation (or overall) emerging as the winner(s). Assuming all of the previous work has been performed in simulation, this dance is then transferred onto the real humanoid robot. Of course some, or all, of the evolutionary process could take place on the real humanoid in order to reduce the likelihood of an inaccurate transfer of the evolved dance from the simulator onto the real humanoid; this is the "reality gap issue", a topic which we discussed in some detail in Chap. 4.

If this process is performed with just one human involved we may evolve a dance suited to his/her specific taste. If it is done with multiple evaluators, a dance more suited to the general observer may be evolved.

5.6.1 An Example Outline Interactive Evolutionary Algorithm for Generating an "Evolutionary Dance"

Algorithm EVOLUTIONOFDANCE

Input: P_c; probability of crossover
P_m; probability of mutation
NoGens; no. of generations for objective phase,
PopSize; population size for objective phase

Output: the best evolved dance

INITIALISE GO(t=1) randomly
; assign random positions within specified limits to all of the
; humanoid robot's joints, over a complete dance sequence
For t = 1 to *NoGens*
    For i=1 to *PopSize*
        OBJECTIVEEVALUATE (individual(i))
    End For
    For j=1 to *PopSize*
        SELECT(GO(t), individual(k))
        Copy individual(k) to chosen_individual
        CROSSOVER(P_c, chosen_individual, random_individual)
        MUTATE(P_m, chosen_individual)
        Copy chosen_individual to GO(t+1) as individual(j)
    End For
End For
Extract 10 best performing dances from GO(*NoGens*+1) creating GS(1)
While user wants to continue the evolutionary process
    t=1
    For i=1 to 10
        SUBJECTIVEEVALUATE (individual(i))
    End For
    For j=1 to 10
        SELECT(GS(t), individual(k))
        Copy individual(k) to chosen_individual
        CROSSOVER(P_c, chosen_individual, random_individual)
        MUTATE(P_m, chosen_individual)
        Copy chosen_individual to GS(t+1) as individual(j)
    End For
    t++
End While

Choose best evolved dance from final generation as the "winner"
Transfer this evolved dance onto the real humanoid robot

Chapter 6
The State of the Art in EHR

6.1 Developments in EHR from 2000 to the Present

It is now over half a decade since my initial short survey of the EHR field entitled "Evolutionary Humanoid Robotics—Past, Present and Future" was published in the book *50 Years of Artificial Intelligence: Essays Dedicated to the 50th Anniversary of Artificial Intelligence* (Lungarella et al. 2007). As its name suggests, this was a publication designed to mark the 50th anniversary of the formal inception of the field of artificial intelligence as a separate research domain, with its own distinct attributes; although there are those who would argue that over half a decade earlier the British scientist Alan Turing was the main protagonist and instigator of this field.

I knew that the field had moved on since this initial foray, however, it was not until a couple of months into this project that I realised quite the extent of that progress. It appears at this stage that every couple of weeks advances are being made in the area of intelligent robotics, and in the field of humanoid robotics in particular. In this chapter we detail some of the major research developments in the EHR field, from the year 2000 to the present. This chapter is divided into two sections: one covering developments in 2000–2007, and a separate section in easily accessible tabular form covering recent research in the period from 2008 to the present day.

6.1.1 The Early Years: 2000–2007

6.1.1.1 Introduction

We initially address what could be termed the "early years" in the EHR field, the years 2000–2007. While this choice is somewhat arbitrary, I picked the year 2000 as the starting point for this period as it was in this year that the textbook

© The Author(s) 2015
M. Eaton, *Evolutionary Humanoid Robotics*,
SpringerBriefs in Intelligent Systems, DOI 10.1007/978-3-662-44599-0_6

Evolutionary Robotics was published by Stefano Nolfi and Dario Floreano, marking, in a sense, the maturation of the field of evolutionary robotics; also in the year 2000 the groundbreaking Honda humanoid robot ASIMO hit the world stage. These years were characterised by the increasing use of more complex robots with higher numbers of degrees of freedom, and the more common use of simulators employing reasonably accurate physics simulators along with more sophisticated graphical representations of the humanoid robots' behaviours in simulation. Much of the work in this period consisted of efforts to synthesise bipedal locomotion using evolutionary techniques.

Again, we note that the categorisation of research work by particular researchers into one time period or another is a little arbitrary; also it would be very difficult to describe every piece of research conducted by all researchers in EHR from 1990 to the present day. However, we attempt to outline significant work(s) by researchers in the area with, where appropriate, references to earlier or later related works by these researchers. For convenience of referencing we divide this analysis into two sections: what we will call embodied experimentation in which either some or all of the evolutionary process was conducted on a real humanoid robot, or where the results of the evolutionary process were tested/transferred to a real humanoid; and simulated experimentation, where no physical humanoid robot was employed in the experiments.

6.1.1.2 Embodied Experimentation

An early piece of research on the evolution of both the morphology and control of a humanoid robot was conducted by Ken Endo and colleagues (Endo et al. 2002, 2003a, b). Some of these experiments involved the evolution of bipedal locomotion for the humanoid robot PINO, using a genetic algorithm to choose the control parameters for simple oscillatory circuits. Ten of the robot's 15 degrees of freedom were utilised in the walking motion, which used a fitness function based on energy consumption and the height that the leg was lifted, among other factors. They also conducted work in simulation to evolve both the controller and morphology of a robot with ten joints, broadly based on the PINO robot. In this case the length of five of the ten links of which the body was constructed was subjected to the evolutionary process, however, the total length of all links remained constant. The idea here was to potentially improve the overall structure of the PINO robot for bipedal locomotion. The authors noted, however, that while stable walking gaits were generated by their method, these gaits bore no resemblance to those exhibited by humans.

Wolff and Nordin conducted some of the earliest experiments involving the application of evolutionary techniques to evolve behaviours directly on a real humanoid robot without the requirement of an intermediate simulation step, and thus effectively almost completely avoided the reality gap issue, which arises when some or all of the evolutionary processes take place in simulation (Wolff and Nordin 2001, 2002). This advantage did not, however, come without an associated cost. They reported the need for continuous maintenance of the robot platform involving the replacement of the knee servos three times each, together with the hip, ankle, and torso motors over a six-month testing period. The robot used in the evolution was the ELVINA robot which, when scaled, has humanlike dimensions. It had a height of 28 cm, and 14-DOF in total, 12 of which were subject to the evolutionary process. A gait was specified by giving the position of each of these 12-DOF for a total of 9 keyframes, together with a speed and a delay parameter for each of the keyframes, giving a total of 126 integer-valued genes per chromosome. By a process of interpolation, the movements between each of these keyframe values were smoothed out to encourage fluid movement patterns. The experimental setup involved the robot being placed on a tabletop with a horizontal beam directly above carrying a power supply cable and a security chain, both of which moved along with the robot. Each population consisted of 30 individuals, 4 of which were randomly picked for evaluation, over a total of 9 generations. The two individuals with higher fitness were chosen as parents, and their children replaced the two individuals with lower fitness. The fitness function used was based on the product of the average velocity of the robot during a trial and a "straightness function". Starting off with a manually seeded population capable of static walking, the authors reported that the evolutionary strategy employed generated straighter and more robust bipedal locomotion.

Wolff and Nordin then moved to a simulation-based approach in which they used the open dynamics engine (ODE) to simulate the ELVINA robot described above, and used linear genetic programming to evolve forward locomotion (Wolff and Nordin 2003a, b). They recognised the difficulties in evolving behaviours purely on the real robot:

> Evolving efficient gaits with real physical hardware is a challenge, and evolving biped gait from first principles is an even more challenging task. It is extremely stressing for the hardware and it is very time consuming.

Their idea was to evolve control programs initially in simulation and then to transfer this evolutionary process to the real robot; however, because of technical issues they were unable to complete the transfer of the evolved behaviours to the real robot. They did succeed in the evolution of forward locomotion of the robot in simulation, while observing many instances of the robot falling over. Since this particular set of experiments was conducted solely in simulation, no damage was done to robot or to the environment!

In 2001 Langdon and Nordin conducted research on the evolution of hand-eye coordination using the 60-cm-tall humanoid robot "Elvis", which contained 42 servos controlling the robot's arm, leg, and hand movements (Langdon and Nordin 2001). The experiment was conducted in two separate parts. In the first part the humanoid robot waved its hand in front of its face, recording the perceived location of its finger tip (which had an attached red laser light source) using the robot's stereo vision system, together with the commands used to drive its hand and arm into position. The evolutionary algorithm employed was the Discipulus machine code genetic programming system which was used to evolve a function to drive each of the servos in the robot's arms. In the second part of the experiment, a stationary target (identified using an identical red laser) was placed in front of the robot, which it then attempted to touch using its evolved function set. The authors claim that the robot developed humanlike performance, and that the results obtained demonstrated the potential of their approach in practical control applications of the humanoid robot. Separate GP based evolutionary experiments were also conducted using the Elvis humanoid robot in the areas of the evolution of stereoscopic vision (Graae et al. 2000) and sound localisation (Karlsson et al. 2000).

In 2003 Zhang and Vadakkepat reported the use of an evolutionary algorithm as an aid in the design of walking and stair climbing in a 12-DOF simulated humanoid robot (Zhang and Vadakkepat 2003). The robot simulated was the RoboSapien humanoid which was designed and fabricated in the Automation lab of the National University of Singapore and which was modelled using the Yobotics simulator. The optimal hip height for walking was determined by examination of human walking, and the zero moment point (ZMP) criterion was used in the determination of the stability of gaits generated. The authors further claim to have implemented the evolved gaits on a real RoboSapien robot.

Liu and Iba proposed an approach they termed "CBR augmented GP" in an application which involved moving a real robot around an environment scattered with obstacles, with the goal of approaching an object, picking it up, and then carrying it to a goal (Liu and Iba 2004a, b). The robot used was the HOAP-1 robot from Fujitsu Inc., which is 48 cm tall. They divided the control system of the robot into two separate parts, a high level planning layer using genetic programming, and a lower-level reactive layer incorporating case-based reasoning (CBR). The idea is that the high-level planning, which is handled by GP, evolves a set of nine abstract behaviours such as move forward (MF), move left (ML), turn right (TR), pick up (PU), etc., using a simple offline 2D mobile robot simulator, where the humanoid robot is treated as a point object. Fitness at this level is based on the ability of the robot to accomplish the task, the number of steps the robot takes, and the total number of collisions. A case base is then used, in conjunction with the information gleaned from the camera with which the robot is equipped to detect similar real-world situations to those previously stored in order to generate a concrete behaviour, based on current environmental conditions, from the abstract behaviours generated by the GA. This second stage takes place on the real robot, and successful object-moving behaviours were generated by this two-stage process.

In 2004 Boeing and his colleagues at the University of Western Australia used a genetic algorithm to evolve bipedal locomotion in simulation using a 10-DOF humanoid model, simulated using the DynaMechs library, for subsequent transfer to the a real 35-cm-tall humanoid robot called Andy (Boeing et al. 2004). This was based on their previous work conducted entirely in simulation on the evolution of walking and jumping behaviours (Boeing and Bräunl 2002). They used a spline-based approach, where the genetic algorithm chooses the control points for three separate splines: a start spline to get the robot from its initial position to the start of the walking motion, a cyclic spline which is repeated as many times as required to make the robot walk, and an end spline to end the walk and move the robot back to a stationary position. Slow forward walks were evolved in simulation; however, when these walks were transferred to the real humanoid robot Andy, not many of the walks transferred successfully, and the resulting locomotion was inferior to a well-constructed manually designed gait.

Kee and colleagues used a genetic algorithm to tune a set of proportional integral (PI) parameters for the real humanoid robot GuRoo (Kee et al. 2004). This robot was developed by the University of Queensland and stands a total of 1.2 m tall, the size of a six year old child, making it one of the larger humanoid robots to which evolutionary techniques have been applied to date. The aim was to achieve robust and stable locomotion, with the eventual aim of creating soccer playing abilities. The robot has a total of 23 joints, 15 of which were subject to the evolutionary process. The number of gains for tuning was also reduced from 30 (15 each for the P and I components) to 18 from considerations of symmetry, each of which was coded by a 2-bit value, giving a total genome length of 36 bits. The evolution was first performed on a simulator based on DynaMechs using a model of GuRoo, for subsequent transfer to the real robot The fitness function used was based on tracking error and smoothness criteria, and the parameters evolved using the GA tuning method were generally superior to those obtained by hand-tuning, especially when transferred to the real robot.

In an approach which has parallels with (Liu and Iba 2004a, b) as described earlier, and which was published in the prestigious journal *IEEE Transactions on Evolutionary Computation*, Kamio and Iba describe a set of experiments involving the combination of genetic programming and reinforcement learning as applied to a box-moving task using both the doglike robot AIBO and the humanoid HOAP-1 robot (Kamio and Iba 2004). As in the previous research, a precise simulator was not required for the first part of this experiment, the GP phase. The robot was represented as a circular object on a two-dimensional surface, and the task of the robot was to push a box onto a particular point on this surface. Three separate actions were possible as a result of the initial GP layer: move forward, turn left and turn right. The fitness function employed was quite a complex function involving issues such as whether the box was moved at least once, whether the robot faced the box or the goal at least once, and whether the assigned task was actually completed. Following this initial work in simulation, reinforcement learning was then used to apply the evolved motion patterns to situations involving real robots—the Sony AIBO robot and the Fujitsu HOAP-1 humanoid robot—using an additional rein-forcement learning layer. They demonstrated in their research that this technique produced superior results to the RL Q-Learning method on its own.

In 2005 Kambayashi and his associates conducted research on the generation of optimal gaits using a genetic algorithm, based on the ZMP criterion (Kambayashi et al. 2005). They claim in their work that this is the first research to directly generate a biped gait using a GA (as opposed to evolving weights of artificial neurons). Fitnesses were computed as the difference between the ZMP as computed by the evolutionary algorithm and the target ZMP. Their application generated bipedal locomotion for a 17-DOF robot, of which 10 of these DOF were subject to evolution. The robot used was the FD Jr. humanoid robot by Best Technology, Inc.

In 2006 Hebbel and his colleagues at the University of Dortmund described a series of experiments designed to produce forward walking in the 17-DOF KHR-1 humanoid robot. These experiments use an evolution strategy approach in simulation using the SimRobot simulator, which in turn is based on the Open Dynamics engine (Hebbel et al. 2006) Interestingly, several different optimisation techniques, including simulated annealing and particle swarm optimisation, were also evaluated in this work.

The final evolutionary process took place autonomously on the real robot using the best optimisation technique discovered using their research, and was used to evolve a forward walk on the real KHR-1 robot. The researchers emphasise that in this phase the robot learned to walk in a completely autonomous fashion—there was no tethered battery back or computational connection involved.

The issue of gait generation was also considered by Capi et al. (2005, 2006) in relation to a five-link bipedal robot considering motion in the saggital plane alone. The parameters chosen for the simulations conducted were based in the "Bonten-Maru I" humanoid robot, which stands 1.2 m high. A multiobjective evolutionary algorithm was employed based on reducing energy consumption and on the minimisation of torque change. The evolved gaits were transferred successfully to the real robot resulting in stable locomotion. Earlier related work in this regard also demonstrated the evolution (in simulation only) of stair climbing based on the parameters of the "Bonten-Maru I" humanoid robot (Capi et al. 2001a, b).

In 2007 Hein and his colleagues at the Humboldt University Berlin described experiments aimed at the generation of robust walking at reasonably high speed with particular application to the RoboCup community (Hein et al. 2007). Their work was conducted in simulation based on an accurate simulation of the Bioloid humanoid robot generated using the Open Dynamics Engine (ODE). The motions were generated using a "neural oscillator approach", where fitness was based primarily on the distance travelled by the robot in a defined direction. Thirty-four synaptic weights were evolved to produce trajectories for 10 of the robot's joints (left- and right-hand sides of the robot were assumed to make the same movements in different phases). The transfer of the evolved motions to the real Bioloid humanoid robot was not particularly successful, however, for reasons outlined by the authors in their paper. These issues involved technical aspects of the robot's operation including the tolerances of gears and the motion characteristics of the servo motors, both of which were not modelled sufficiently in the simulation in order to allow for successful direct transfer of evolved gaits from simulation to the real Bioloid robot. It turned out that all of the evolved motions required manual tuning in order to stabilise the robot, again illustrating the prevalence of the "reality gap" issue.

Also in 2007, a genetic algorithm was employed for both walking and stair climbing based on a model of an eight-DOF robot using MATLAB to model the joint angles (Shrivastava et al. 2007). Their model used cubic spline interpolation in the generation of robot motions, which involved computing the joint angles required to maintain balance in the motions using the ZMP criterion and attempting to minimise energy consumption. Unusually for the experiments described here, a deformation of the sole of the humanoid robot's foot was allowed, and a method of correcting for this deformation was proposed. They used a chromosome consisting of 16 genes providing the values of 2 control points for each spline for each of the robot's 8 joints. Four control points per spline were used, and the starting and ending points of each spline were predetermined. The genetic algorithm was in charge of the generation of two intermediate control points, which were coded as real variables. It was shown in this research that by correcting for sole deformation the total energy consumed was minimised, demonstrating the possible useful application of their proposed method of correcting for sole deformation.

Kulvanit et al. (2007) demonstrated the application of evolutionary algorithms to the generation of fast biped walking using a fitness function based on walking speed, the stability of the robot based on the amount of front-back sway generated, and the power consumption. They applied their work to adapting the parameters of a real humanoid robot called Jeed that has a total of 22-DOF and stands 45 cm tall, and which was designed to compete in the RoboCup humanoid league soccer competition. Starting from the best human-designed gait, the genetic algorithm is then allowed to optimise four separate parameters defining the structure of this gait. These are: the step length, the side sway distance, the angle of the trunk bend, and a time constant based on the period between a foot being lifted from the ground to when it is placed on the ground again. Each of these parameters is represented in the chromosome as a 10-bit value, giving a total chromosome length of 40 bits. The model is generated in simulation for transferral onto the real humanoid robot.

6.1.1.3 Simulated Experimentation

We now turn our attention to experiments in the period 2000–2007 involving evolutionary algorithms applied to simulations of humanoid robots, without being actually applied to real robots. In 2001 Ogihara and Yamazaki constructed a sophisticated 7-link neuro-musculoskeletal model based on the lower part of the human body, and which was expressed as 76 separate simultaneous equations (Ogihara and Yamazaki 2001). The genetic algorithm in this instance was used to tune 93 neural parameters based on this model for the generation of bipedal loco-motion. The fitness function employed was based on the minimisation of energy expenditure, and numerical simulation was employed using the Euler method. Walking behaviour was produced eventually by this method, but reproducibility of the results generated was not a given in these experiments.

In 2001 Jimmy Pettersson and his colleagues at the Chalmers University of Technology performed research on bipedal locomotion involving a simulated five-link robot with five degrees of freedom which was constrained to move in the sagittal plane alone (Pettersson et al. 2001). Interestingly, in this paper the researchers used the term "robotic brain" rather than "control system" to describe the controller for the humanoid robot to emphasise the greater influence that the program generated had on the actions and abilities of the robot as opposed to a controller employing classical control theory. In their own words.

> … we will use the term robotic brain for the computer program that determines the actions of the robot, rather than the term control system. The latter term would indicate a more limited representation employing classical control theory.

They used a technique based on one of the earliest EAs, as espoused by Fogel (1966), which used an evolutionary algorithm to evolve a finite state machine (FSM), in a process known as evolutionary programming. The results obtained were mixed, but illustrated the utility of continuing with such a research approach.

Again in 2001, Fujii et al. conducted research on a controller for a seven-DOF biped robot (in simulation only) using central pattern generators (CPGs, Fujii et al. 2001). They used the MathEngine simulation package to simulate a seven-DOF biped (lower limbs only) which was controlled using a dynamically reconfigurable neural network (DRNN) controller consisting of a set of neural oscillators. The fitness function was based on the distance travelled in the trial period (30 s), without the robot falling down, taking into consideration the distance travelled by both the left and the right foot to encourage locomotion involving alternate steps by both feet.

The issue of bipedal locomotion was again addressed by Bongard and Paul at the University of Zurich using a five-link six-DOF model of the lower body of a humanoid using a physically realistic simulation also based on the MathEngine simulation package (Bongard and Paul 2001). Interestingly, they also allowed the inclusion of morphological considerations in some of the experiments. A genetic algorithm was used to evolve the weights for a recurrent neural network with 60 synapses. In the experiments which allowed for morphological change an additional three or eight parameters are encoded in the genome, which are distributed evenly across the genome to increase the recombination possibilities for these parameters in crossover. These parameters encode factors such as the radii of different segments of the simulated robot's body, and the lengths and positions of the mass blocks which are incorporated in the robot's limbs. Inputs to the network included touch sensors in the feet, and proprioceptive sensors in each of the six joints. Fitness is based on the distance travelled in the northern direction (the direction the robot faces initially), either at the end of a set simulation period, or, in the event of premature termination of an evaluation, on the distance travelled up to this termination. The researchers conclude that "the arbitrary inclusion of morphological parameters does not always yield better results".

In a separate but related piece of research, three experiments were conducted allowing the mass blocks to represent different fractions of the total weight of the biped (Paul and Bongard 2001). Based on these experiments, which they claimed demonstrated for the first time that stable locomotion for biped robots could be achieved by coupling the optimisation of both the robot's morphology and control, the researchers concluded that "including morphology in the optimization process can lead to much more efficient design in some cases".

In 2002 Ishiguro and colleagues at Nagoya University described a set of interesting experiments which they considered to be the first to evolve locomotion controllers that could create passive-dynamic walking (Ishiguro et al. 2002). In this work they investigated the mutual interaction between controllers and their environment. The body parameters of a four-joint simulated biped robot were evolved to create passive-dynamic walking. A two-stage evolutionary process was employed; first the physical (body) parameters of a simulated bipedal robot were evolved such that it supported passive-dynamic walking. Second a controller was evolved for the passive-dynamic walker that was evolved in the first stage, such that energy consumption was minimised and distance travelled was maximised. Further experiments were then conducted on the evolution of similar controllers for morphologies which did not support, or barely supported, passive-dynamic walking. It was demonstrated that controllers evolved for robots in which the robot morphologies were initially evolved to support passive-dynamic walking considerably outperformed those of these latter experiments, thus demonstrating the increased evolvability of these embodiments.

In 2002 Wakaki and colleagues in the University of Tokyo used interactive evolutionary computation (IEC) to create motions for a 3D computer graphics (CG) avatar modelled using the Humanoid Animation Standard (H-Anim) (Wakaki et al. 2002). The range of motions displayed on the screen was limited to those that could be accomplished by an actual human. Genetic programming was used as the underlying evolutionary algorithm, and motions such as "dancing in time to the given music" and bowing were evolved.

In a landmark paper published in 2002 in the prestigious publication, the *IEEE Transactions on Evolutionary Computation*, Reil and Husbands demonstrated the evolution of bipedal locomotion in a six-DOF humanoid model using the Math-Engine simulator, as in Bongard and Paul's work (Reil and Husbands 2002). Their approach was to evolve the weights, and some other parameters for recurrent neural networks to generate stable bipedal locomotion for their physically simulated biped robot. They claimed that,

> to our knowledge, this is the first work to demonstrate the application of evolutionary optimisation to three-dimensional physically simulated biped locomotion

The fitness function employed was designed to maximise the distance travelled by the simulated robot, while not allowing its centre of gravity to fall below a certain height. No crossover was employed, and rank-based selection was used, where the bottom half of the population (in fitness terms) was removed and replaced with a second copy of the top half. Stable walking was developed in simulation using their approach; however, the authors acknowledged the potential problems in transferring their results evolved in simulation onto a real robot because of possible inaccuracies in the simulator employed. Experiments were also conducted, with mixed success, in evolving walking in the direction of a particular sound source.

The researchers noted the high gait diversity in the walks generated, and the similarity to human walks in some instances, although this was not selected for specifically in the evolutionary process. It should, however, be noted that only 10 % of the evolutionary runs ended up in a stable walking gait.

In 2003 Miyashita, Ok, and Hase demonstrated the use of genetic programming in order to evolve biped walking for a 12-segment humanoid model (Miyashita et al. 2003). This article is related to an earlier 2001 article on this subject (Ok et al. 2001). Unusually for most of the experiments described in this section, the upper part of the simulated humanoid was also modelled, including the upper part of the torso and the arms. The lil-gp genetic programming system was used as their simulation platform, and fitness was evaluated based on a linear combination of the distance the simulated humanoid walked before falling down, the number of steps made while walking, and the prevalence of horizontal and vertical shaking motions of the robot's body. The genetic program was used to determine the parameter values for 8 neural oscillators working as central pattern generator; the number of terminals used by the GP was 121, a large number by normal GP standards. The number of feedback structures used was reduced from 12 to 8 because of symmetry considerations involving the left–right symmetry of the human body which affected 8 of these DOF. While humanlike bipedal locomotion did evolve, this was limited to ten steps because of the inherent instability of the limit cycles generated.

Also in 2003, Hase, Miyashita, Ok, and Arakawa demonstrated the simulation of human gait using a precise neuro-musculoskeletal model in an interesting paper which employed evolutionary computation to generate a variety of humanoid locomotive patterns (Hase et al. 2003). These included normal gait, pathological gaits (i.e., a person with an artificial limb), running, and apelike walking. They used a 19-DOF model of the human body incorporating 14 rigid 3D bodies and 60 muscular models. The genetic algorithm was used to evolve 125 parameters of the underlying neuronal control system. Different fitness functions were used in the evolution of the different behaviours; in the experiments designed to emulate the evolution from ape to human locomotion, additional morphological characteristics were added to the GA search. A model of a chimpanzee was first constructed, and the GA was then applied to this evolved chimp model and their results obtained compared to both the body shape and the locomotive patterns of modern humans. While not exactly matching the morphological structure or walking patterns of modern humans, the authors suggest that their work should provide a useful foundation for further anthropological studies of human evolution and that their work could prove useful in computer graphical applications involving simulated humanoids. Their hope also was that this work could form a connection of sorts between human biomechanical research and computer animations.

Ishiguro et al. addressed the issue of bipedal locomotion over flat and inclined terrain using a two-stage evolutionary approach (Ishiguro et al. 2003). They used MathEngine to simulate the lower body of a seven-DOF humanoid robot. Seven sets of neural oscillators were employed, the parameters of which were evolved by a genetic algorithm. A process of incremental evolution was employed, and the fitness in the first stage was based on the period travelled by the robot during the simulation period (30 s). The second stage of evolution used as its initial population the last generation of the previous stage, and the simulated robots used a fitness based on the product of the distance travelled on the uphill and downhill slopes, respectively. Successful locomotion on all three terrains (flat, uphill, and downhill) were evolved using this approach.

In 2004 Lee, Kim, and Lee from KAIST, Korea demonstrated the evolution of bipedal locomotion for a 6-DOF simulated biped robot using 12 parameters in total for the evolutionary process (Lee et al. 2004). Quite a complex fitness function was employed, taking into account considerations such as the stability of the robot, energy considerations, and a variety of basic constraint conditions. Useful walking trajectories were obtained using their methodology.

Also in 2004 Jeon, Kwon, and Park of Hanyang University, Korea demonstrated the use of genetic algorithms for staircase climbing in the case of a six-DOF simulated humanoid (Jeon et al. 2004). They used MATLAB to evolve the loco-motion of the robot using the consideration of energy optimisation. A number of constraint considerations were also taken into account, including the conditions of the stairs, the knee joints of the robot, and ZMP conditions. Satisfactory staircase climbing behaviour was demonstrated in their experiments.

At the University of Sussex in 2004 Vaughan et al. demonstrated the evolution of bipedal locomotion in a 3D ten-DOF powered passive dynamic robot (Vaughan et al. 2004). They used the ODE for their experiments. The speed of the evolved robot could be dynamically adjusted, and it was capable of adjusting to different environmental conditions and changes in its own morphology. The fitness function employed was based on the selection of robots that move as far as possible in a straight line. They did this with what they called a subsumption approach, which would now be called incremental evolution, where they first evolved feed-forward continuous-time neural networks without sensory feedback, and later added sensory input to the evolutionary process. Using this approach the evolution of robust walking behaviour was observed. In another research article in 2005, these researchers demonstrated the use of a genetic algorithm to evolve 12 parameters for a six-DOF simulated humanoid robot, including morphological characteristics (Vaughan et al. 2005). These evolved robots demonstrated the ability to adapt their gaits in a dynamic fashion in response to external forces and exhibited robustness to noise interference.

Park and Choi also used a six-DOF simulated humanoid in their research, which aimed at minimising the energy consumption of a biped robot in locomotion (Park and Choi 2004). This involved another early effort to evolve both the body and the brain of the simulated robot, as a GA was used to search for the optimal positions of the mass centres in the links which held the robot together (all of the mass of each link is assumed to be concentrated on a single spot) as well as evolving optimal trajectories for the robot's legs.

In 2004, McHale and Husbands conducted research comparing variations of 3 different types of neural network (14 different ANNs in all); the continuous time neural network (CTRNN), the plastic neural network (PNN), and the GasNet, all of which were as part of an evolutionary process applied to bipedal locomotion for a

5-DOF simulated robot which was "not entirely physically realistic" (McHale and Husbands 2004). Their research suggested the potential utility of GasNets in solving future sensorimotor control problems of this nature.

In 2005 Tang, Zhou, and Sun presented a two-stage approach to the generation of efficient walking using GAs (Tang et al. 2005). The first stage involved the generation of walking gaits based on the ZMP criterion and the energy consumption of the four-DOF simulated robot. The second stage employed a genetically trained neural network to generalise the walking gaits evolved in the first phase. Using this two-stage process the researchers claim the generation of near-optimal walking gaits based on varying step lengths and walking cycles.

In 2005 Alankus and colleagues presented research involving the generation of dancing motions for a synthesised humanoid character based on analysing the musical beats of a song or of a melody (Alankus et al. 2005). They claim in their research to be, to the best of their knowledge,

> the first researchers who have created a fully automated dance sequence generator of human animation

Their research involved the use of a stock motion library containing motion capture data and had the ability to generate novel motion sequences not contained in this library. They used both a greedy algorithm and a genetic algorithm to aid in the task of motion synthesis, using the same evaluation (fitness) function. This fitness function was based mainly on the level of synchronisation of the music beats with the dance moves of the simulated robot. They noted the greedy algorithm to be faster than the GA; however, in situations where the variety of figures was of importance or the beat intervals of the song fluctuated at a high rate, the genetic algorithm approach could be superior.

Chen and colleagues from the Hebei University of Technology proposed a methodology in 2007 for optimisation of the gaits of a 12-DOF simulated humanoid. The robot was simulated using MATLAB, and it involved mixing both binary and floating-point representations in the genetic algorithm (hence the "mix" in the paper's title). Walking stability (based on the ZMP criterion) and energy optimisation were two of the criteria considered as part of their analysis (Chen et al. 2007).

In 2007 Ha, Han, and Hahn discussed the issue of the synthesis of a bipedal gait based on the analysis of a human's gait (Ha et al. 2007). A genetic algorithm was used to adaptively create a gait pattern for a five-link simulated humanoid in the sagittal plane; additionally a gait pattern for the frontal plane was evolved based on minimising the distance between the robot's ZMP value and a theoretically calculated ZMP value. The evolved gait was simulated using OpenGL and was shown to closely approximate that of the human modelled.

Jingdong et al. (2007) from the Harbin Institute of Technology discussed the improvement of walking ability for a six-DOF simulated robot in the context of robot soccer playing skills for the Federation of International Robot-soccer Association (FIRA) competition, in which their robot had had previous success. They also addressed the generation of penalty kicking and goalkeeping behaviours in this paper; however, evolutionary techniques are not employed in these cases. The optimal ZMP trajectory was computed for this robot based on four separate critical parameters.

Also in 2007 Heralić, Wolff and Wahde conducted research into the evolution of central pattern generators (CPGs) for gait generation in bipedal robots (Heralić et al. 2007). They compared the results obtained between two different approaches. Method 1 involved evolution in two stages; the first stage involved the simulated robot using a posture support mechanism, with several joints locked in place. Once a stable individual was obtained this was cloned in order to create individuals for the second phase, where the support structure was to be removed, and most joints unlocked. Method 2 involved a one-step evolution process, with most joints unlocked throughout the evolutionary process. Although the original formulation involved Method 1 having the support structure in place in the first phase, and removed in the second, this was altered to allow the positive support to be in place for the first two seconds of experimentation in both phases. This approach was also adopted for Method 2. The simulated robot had either 8 or 12 degrees of freedom and the fitness function employed was broadly based on the distance travelled by the robot in the forward direction reduced by deviation in the sideways direction.

Finally, in 2007 Yang and colleagues discussed a methodology that used a genetic algorithm to generate the coefficients of a truncated Fourier series (TFS) for the generation of stable bipedal gaits on both flat and inclined surfaces using a technique they term the genetic algorithm-optimised Fourier series formulation (GAOFSF, Yang et al. 2007). Their research was applied to a seven-link simulated planar biped robot, and stable walking was demonstrated on both flat and inclined surfaces. The simulation environment used was Yobotics. The fitness function for the genetic algorithm was based on ZMP stability considerations, together with six separate penalty functions to encourage correct and natural-looking walking patterns. The authors claim that an advantage of their approach is the ease with which the both the length and frequency of the stride can be adjusted.

6.2 Recent Research Developments in the Field of Evolutionary Humanoid Robotics

In this section we present a review of the current state of the art in of the field of EHR over the last six years or so, that is, since the year 2008. This research is categorised in several fashions, including the general application area, and the type of evolutionary algorithm employed. Of course, any survey of this sort will be incomplete by nature given the ongoing dynamic nature of research being conducted, however, we hope that it is possible to give a snapshot of some of the details of research in most of the broad areas of research in this field. The overall results from this survey are presented in Tables 6.1 and 6.2, covering simulated and embodied experimentation respectively; each table is organised by year of publication in reverse chronological

Table 6.1 Simulated experimentation

Year	Researchers	Platform-simulator	Application	Fitness function type	Type of evolution	What is evolved
2013	Al Borno, de Lasa, and Hertzmann	Simulation of full body motion using Featherstone's algorithm	Synthesising a wide range of human movements, including walking, crawling and breakdancing	Various tailored functions for the different movements	Covariance matrix adaptation (CMA)	350 variable values representing key-poses of a cubic-B-spline parametrised by Euler angles
2013	Santos	Open dynamics engine (ODE)	Bipedal locomotion for a six-DOF simulated robot on flat surfaces, on a slope, and on stairs. Compare results obtained for synaptic delay neural networks with those for continuous time neural networks based on Santos and Campo (2012)	Technically tailored but mainly aggregate—based on distance traveled by the biped in a given time (8 s)	Genetic algorithm	A varying number of parameters for both synaptic delay neural networks and continuous time recurrent neural networks
2012/ 2008	Wang, Lu and Zhang	Not specified	Motion generation (ladder climbing) for 23-DOF simulated humanoid robot	Tailored	Genetic algorithm, using MATLAB genetic algorithm toolbox	Generation of keyframes
2012	Jadhav, Joshi and Pawar	N/A	Generation of novel dance steps for an Indian classical dance (BharataNatyam) for choreographer use	Tailored	Genetic algorithm	Dance steps represented as dance vectors
2012	Ouannes, Djedi, Duthen and Luga	Open dynamics engine (ODE)	Bipedal locomotion, 15-DOF	Aggregate	Genetic algorithm	Recurrent neural network
2012	Wang, Hammer, Delp and Koltun	Open dynamics Engine (ODE)	Automatic synthesis of controllers for walking and running for a simulated 30-DOF humanoid character representing a male adult	Tailored	Covariance matrix adaptation (CMA)	124 parameters defining a simulated motion

(continued)

Table 6.1 (continued)

Year	Researchers	Platform-simulator	Application	Fitness function type	Type of evolution	What is evolved
2012	Torres and Garrido	Webots	Automatic synthesis of walking behaviours for a model of the Aldebaran Robotics Nao humanoid robot, for 6-DOF, and compare results with two other walk engines	Aggregate (?)— precise fitness function not specified	Genetic algorithm	16 parameters for 6 neural oscillators controlling the robot's motion
2012	Savastano and Nolfi	Custom simulator, available online: (http://laral.istc.cnr.it/laral++/farsa), based on Newton Game Dynamics open-source physics engine	Reaching and grasping tasks, modelling those of human infants, applied to a 14-DOF model of the iCub humanoid robot	Tailored, incremental	An "evolutionary method" based on GAs	Neural network based controller
2011	Lehman and Stanley	Open dynamics engine (ODE)	Bipedal locomotion. 6-DOF	"Novelty search" versus aggregate noninteractive	NeuroEvolution of augmenting topologies (NEAT)	Neural network
2011	Azarbadegan, Broz, and Nehaniv	Simulator adapted from Thomas Miconi's work, which is based on Karl Sims' original simulator; uses open dynamics engine (ODE)	Evolution of bipedalism	Tailored	Genetic algorithm	Array of genes representing morphology and connectivity of individual limbs, together with neural information for a McCulloch-Pitts based neural controller
2011	Cardenas-Maciel, Castillo and Aguilar	Mathematical model	Bipedal locomotion minimizing energy consumption of simulated 3-DOF planar biped robot	Tailored	Genetic algorithm	Eight parameters of a feedback controller to produce walks with low energy control

(continued)

Table 6.1 (continued)

Year	Researchers	Platform-simulator	Application	Fitness function type	Type of evolution	What is evolved
2011	Schreiner and Punzengruber	Webots version 6.2.4 modelling 21-DOF Aldebaran Nao robot	To optimise 6 general-purpose motion controllers supplied by the robot's manufacturer	Tailored	Genetic algorithm	35 walk controller parameters for each of 6 motions
2011	Urieli, MacAlpine, Kalyankrishna, Bentor and Stone	SimSpark multiagent system simulator. Uses ODE (open dynamics engine)	Robot soccer skills for simulated Nao robot (22-DOF)—including multi-directional walking, turning, kicking	Tailored	CMA/ES (covariance matrix adaptation evolution strategy) and GA (also hill climbing and cross-entropy method (CEM))	Parameter values for templates which define different skill sets
2010	Sellers, Pataky Caravaggi and Crompton	OpenDE version GaitSym simulator using anybody research project Leg3D model from Model Repository 6.1	Using evolutionary robotics experiments demonstrate the importance of the Achilles tendon, and generate movement patterns for humanoids to predict aspects of human locomotor mechanics, using an 18-DOF model	Tailored, incremental	Genetic algorithm	18×3 parameters in the range -1 to $+1$ representing duration and activation levels for each of the 18 muscles (DOF) for first half of a gait cycle
2010	Dubbin and Stanley	Panda3D simulator	To train virtual humans to dance, 34 DOF in total	Interactive evolutionary computation (IEC)	Approach based on NeuroEvolution of Augmenting technologies approach (NEAT)	ANN, encoding the actuation of the joints of the model

(continued)

Table 6.1 (continued)

Year	Researchers	Platform-simulator	Application	Fitness function type	Type of evolution	What is evolved
2010	Shafii, Aslani, Nezami and Shiry	Resserver3d simulator, based on Spark and ODE	Evolution of straight, stable bipedal locomotion. To compare EA and PSO approaches to walking using simulated Nao humanoid robot with 22-DOF	Aggregate (distance travelled in desired direction)	Genetic algorithm, also Adaptive PSO (both used separately to tune TFS gait, and results compared)	Seven parameters for Truncated Fourier Series gait generator
2010	Wu and Popović	NVIDIA PhysX SDK in software mode	Generate bipedal locomotion which can adapt to uneven ground conditions in real time, for graphics applications, and possible game applications	Tailored, based on energetic cost, COM deviation error, and frame tracking errors	CMA/ES "used as a black box optimizer"	Between 46 and 70 parameters which are normalised to [0,1], initialised to 0.5 to start with
2010	Tuci, Massera and Nolfi	Newton game dynamics library simulating an anthropomorphic robot arm with 27-DOF	To categorise spherical and elliptical objects placed on a flat surface using a robotic arm equipped with tactile sensors	Tailored, based on the sum of two different fitness components	Simple genetic algorithm	Connection weights and other parameters of a recurrent neural network with 22 sensory neurons, 8 internal neurons, and 18 motor neurons consisting of 420 parameters in total, each encoded as 16 bits
2009a	Allen and Faloutsos	Not specified	Generate bipedal locomotion using a neural controller, taking in 11 "sensory inputs"	Tailored	NeuroEvolution of Augmenting technologies (NEAT)	Structure and connection weights of an artificial neural network, which specifies the target angles of seven joints
2009	Sheng, Huaquing, Qifeng and Xijing	Mathematical model of 10-DOF humanoid using Lagrange's equation using MATLAB and GAOT toolbox	Stair climbing gait optimisation based on ZMP (for stability) and minimisation of energy consumption	Tailored	Multiobjective genetic algorithm	Optimisation of ten parameters

(continued)

Table 6.1 (continued)

Year	Researchers	Platform-simulator	Application	Fitness function type	Type of evolution	What is evolved
2009	Suzuki, Gritti and Floreano	Webots model of HOAP-2 humanoid robot	Bipedal locomotion (walking towards a goal)	Aggregate (based on ability of robot to reach goal position)	Genetic algorithm	147 neural network connections, each encoded as five bits (total 735 bits)
2009	Hong, Kim and Kim	Webots simulation of KAIST humanoid robot HSR-VIII (26-DOF)	Footstep planner for a humanoid robot in the presence of obstacles	Tailored—based on number of steps taken, avoidance of obstacles, and minimum energy consumption	Genetic algorithm	24 parameters providing input to univector fields
2009	Kim, Kim and Kim	Model of 7-DOF biped robot based on Euler-Lagrange equation	Walking up and down a staircase	Tailored	Adaptive genetic algorithm (AGA)	Optimisation of six parameters
2009	Vircíková and Sincák	Webots	Evolution of dance choreography for the Aldebaran robotics Nao humanoid robot	Interactive evolutionary computation (IEC), 20 participant evaluators	Genetic algorithm	Angular positions of robots joints
2009	Wampler and Popović	Newton–Euler formulation used for dynamical constraints; SNOPT used for spacetime optimisation	Generate gaits and morphologies for a variety of simulated legged creatures, including a 16-DOF biped	Tailored, based on muscular exertion, head position and stability and the avoidance of high velocity joint motions	Space–time optimisation combined with a variant of CMA	Various, depending on optimisation problem

(continued)

Table 6.1 (continued)

Year	Researchers	Platform-simulator	Application	Fitness function type	Type of evolution	What is evolved
2008b	Yanase and Iba	Simulated HOAP-1 robot, simulator not specified	Footstep planning application, to reduce the number of evaluations required by A* search	Tailored	Non-dominated Sorting genetic algorithm-II (NGSA-II)	Set of parameters for footstep planner: variables representing length of footstep in two dimensions
2008	Boeing	Physics abstraction layer with Dynamechs, ODE and bullet physics library as back-end simulators	Generate biped walking and jumping gaits for two simulated bipeds, a simple biped (6-DOF) and an android biped (7-DOF), also a tripod and a snake	Functional incremental	Genetic algorithm	6(DOF) × 6(control points) 8 bit values for simple biped, 7 × 6 values for android biped + 3 × 16 bit values determining PID parameters (evolved separately)
2008	Zamini, Farzad, Saboori, Rouhani, Naghibzadeh and Fard	Spark, based on ODE, running a model of HOAP-2 humanoid, based on analysis of HOAP-2 gait in Webots	Generate biped walking in a simulated model of the 25-DOF Fujitsu HOAP-2 robot, 12 joints subject to evolution	Aggregate (distance travelled by time passed); tailored function for second experiment	Genetic algorithm	Seven coefficients of a Fourier series, for each of 12 joints = 84 values; second experiment 7 × 4 joints = 28 values
2008	Berger, Amor, Vogt and Jung	Freiberg robot simulator (FRS), a simulator designed to support kinesthetic user input	Use kinesthetic bootstrapping to evolve new behaviours based on initial human-supplied postures for a simulated 18-DOF bioloid humanoid robot	Not specified, dependant on the task to be solved	Genetic algorithm	12 real-coded genes each representing the coordinates of a Bezier control point in the robot posture space

Table 6.2 Embodied experimentation

Year	Researchers	Platform-robot	Platform-simulator	Application	Fitness function type	Type of evolution	What is evolved
2013	Eaton	Robotis Bioloid humanoid (18-DOF humanoid robot)	Webots	Dance	Tailored	Genetic algorithm	Keyframe interpolator, 400 bits–16 bits for each of 18 DOF for 4 keyframes + 2 unused DOF + 4 × 16 bits for keyframe speed + one 16-bit value representing joint range
2012/ 2008	Baydin	Custom hardware involving a lateral boom rotating around a pivot	Mathematical model only involving integration of Newton–Euler equations	Control of a 4-DOF mechanism for bipedal locomotion	Aggregate (total distance moved) although additional tailored functions were tried but not used as minimal improvements in performance were noted	Genetic algorithm	25 real numbers encoding parameters of the central pattern generator network, and topology of the network
2012	Fukunaga, Hiruma, Komiya and Iba	Fujitsu frontech enon service robot, wheeled	Low-fidelity custom simulator	Exhibition guide robot	Tailored	Genetic programming, evolved controller transferred to real robot	Controller program

(continued)

Table 6.2 (continued)

Year	Researchers	Platform-robot	Platform-simulator	Application	Fitness function type	Type of evolution	What is evolved
2012	Sakai, Kanoh and Nakamura	Fujitsu HOAP-1 (20-DOF), and Kondo KHR-2HV (17-DOF) humanoid robots	Open dynamics engine (ODE)	Evolution of "standing-up" motion, both in simulation, and on the real robots, comparing the results of two different approaches	Behavioural—fitness is computed as the sum of the values for the robot's chest position over each individual trial	Evolutionary multivalued-decision diagrams (EMDDs) and evolutionary multiterminal binary decision diagrams (EMTBDDs)	Variable values, and structure of multivalued decision diagrams, and multiterminal binary decision trees
2011	Domingues, Lau, Pimentel, Shafii, Reis and Neves	Aldebaran robotics Nao humanoid robot	SimSpark simulator using ODE	Generate walk behaviour (TFSWalk behaviour)	Not specified	Partial Fourier series optimised with genetic algorithms, PSO and truncated Fourier series also employed	Trajectory of robot joints
2011	Virčíková and Sinčák	Aldebaran robotics Nao humanoid robot	N/A	Dance	Interactive evolutionary computation (IEC)	Genetic algorithm	Direct joint angles
2011	Gökçe and Akin	Aldebaran robotics Nao humanoid robot	Webots	Omni-directional walking for RoboCup applications	Tailored/aggregate	Evolutionary strategy	Thirteen parameters which determine the walking pattern for a CPG based algorithm
2011	Kulk and Welsh	Aldebaran robotics Nao humanoid robot	Webots, using existing RobotStadium environment	Humanoid robot walking, comparing 3 different learning algorithms	Various; tailored	Evolutionary hill climbing with Line Search (EHCLS), together with Guassian particle swarm optimisation (GPSO) and policy gradient reinforcement learning (PGRL)	Two different parameter spaces

(continued)

Table 6.2 (continued)

Year	Researchers	Platform-robot	Platform-simulator	Application	Fitness function type	Type of evolution	What is evolved
2010	Lee, Jong and Yang	Not specified in paper ("a real humanoid robot")	Custom simulation software	Evolve behaviour sequences	Training/tailored, incremental. Imitation-based learning	Modified genetic algorithm	Motor rotation angles
2010	Hettiarachichi and Iba	Fujitsu HOAP-2 robot	Webots	Evolution of 2 asymmetric yoga motions, 11 and 14 joints (DOF) respectively	Tailored, based on humanoid stability, and closeness to the desired pose	Genetic algorithm. Motions evolved in simulation, and transferred to real robot	Joint angle differentials
2010	Kim, de Silva and Park	Custom robot, 19-DOF in total but ten used in experimentation	Not specified	Walking on the level, ascending and descending	Training	Genetic algorithm	Various parameters of Sugeno-type fuzzy models
2009	Antonelli, Dalla Libera, Menegatti, Minato and Ishiguro	Kondo KHR-2HV	USARSim	Generation of humanoid robot movements—walking	Tailored, interactive on one level, user specifies range of permissible joint angles on a semi-interactive basis	Genetic algorithm	10 parameters, 9 representing joint angles, one gives the time between frames
2009	Dip, Prahlad and Kien	12-DOF custom aluminium biped robot using dynamixel DX-113 motors	Mathematical model using Runge-Kutta fourth-order numerical integration and the MATLAB/ Simulink environment	Generation of walking gaits based on stability and walking speed	Tailored	Genetic algorithm	Optimisation of 4 walking parameters based on maximising stability and walking speed

(continued)

Table 6.2 (continued)

Year	Researchers	Platform-robot	Platform-simulator	Application	Fitness function type	Type of evolution	What is evolved
2009	Amor, Berger, Vogt and Jung	Robotis Bioloid humanoid robot (18-DOF)	Physics-based simulator based on open dynamics engine (ODE)	Generate a variety of behaviours including performing a headstand, walking, and standing up, based on initial human-supplied postures	Tailored	Genetic algorithm	Optimise a (variable, depending on application) number of control points specifying a spline curve, which is a compressed representation of the human-supplied motion
2009	Palmer, Miller and Blackwell	Dexter, a 12-DOF adult-size bipedal robot by Anybots Inc. utilising pneumatic actuators	Custom simulator using the open dynamics engine (ODE)	Generation of bipedal locomotion (in simulation only); generation of standing balance for the robot in both simulation and in reality	Tailored, mainly based on time until loss of balance with error terms for variations of various motion values from expected values	Genetic algorithm	22 chromosomes containing the weights for 22 neural networks of four different topological types
2008b	Eaton	Robotis Bioloid humanoid robot (18-DOF)	Webots simulating 20-DOF QRIO-like robot and 18-DOF bioloid humanoid	Generate bipedal locomotion for normal, low-friction, and reduced-gravity conditions. Investigate the effect of morphological constraints	Tailored	Genetic algorithm	336 bits (for simulated QRIO-like robot) representing the value of 20 motors over 4 keyframes together with a joint range value; 400 bits for simulated Bioloid humanoid (includes keyframe speed values)

(continued)

Table 6.2 (continued)

Year	Researchers	Platform-robot	Platform-simulator	Application	Fitness function type	Type of evolution	What is evolved
2008	Wolff, Sandberg and Wahde	17-DOF robot manufactured by Kondo Kagaku Co Ltd. (KHR-1/KHR-2HV)	N/A—experiments on real robot only (fully embodied experimentation)	Optimise the gait for a bipedal robot given an initial hand-coded gait	Tailored	GA without crossover and allowing structural mutations	A variable number of values, representing the joint angles of 13 of the 17 motors (M) at each of up to 10 states (S)—plus a single value encoding the speed of transition between states—total of $14 \times S$ values
2008	Ra, Park, Kim and You	Humanoid robot "Mahru"	Mathematical models	Generate ball-catching type movement	Tailored	EA using a gradient-based local optimisation algorithm	Not clear from article
2008a	Yanase and Iba	Fujitsu HOAP-1 robot	OpenHRP (Open architecture humanoid robotics platform)	Generate various motions for humanoid robots including cooperative dance and kicking using IEC. Optimise motions generated using conventional GA	Aggregate (total distance moved) although additional tailored functions were tried but not used as minimal improvements in performance were noted	Genetic algorithm	41 real-valued parameters including values of 7 DOF for 5 keyframes and time taken per keyframe

order. Some researchers have published similar results over more than one year, and in this case we may give the years involved.

One interesting fact that emerges from this survey is the continued prevalence of the application of evolutionary robotics to the area of bipedal locomotion in humanoids. Over half of the research surveyed involved the evolution of stable walking, at least as part of the published work. Another major application domain is in the evolution of dance behaviours, including dance choreography. Some of this research involves the use of interactive evolutionary computation (IEC) techniques (Takagi 2001). This evolutionary technique essentially involves taking the objective fitness function used by the standard GA and replacing this with a subjective human evaluation of the quality of the evolved behaviours. See Chap. 3 for a further discussion of this issue. A number of behaviours involving robot soccer skills were also evolved. This is not surprising, as a lot of current applications of humanoid robot skills are in the robot soccer domain, specifically in the RoboCup standard platform league, which since 2008 has employed the Aldebaran Robotics Nao Humanoid robot (replacing the Sony AIBO dog-like robot, which was used previously). Other application domains include ball catching, ladder climbing, footstep planning, and crawling and jumping behaviours. Research is also being conducted on the evolution of yoga-like motions, and in the prediction of certain aspects of human locomotor mechanics.

As regards the evolutionary algorithms employed, interestingly over 50 % used a genetic algorithm of some form. Other algorithms used were the neuroevolution of augmenting technologies (NEAT) approach (Stanley and Miikkulainen 2002), the covariance matrix adaptation (CMA) approach (Hansen and Ostermeier 2001), the nondominated sorting genetic algorithm II (NSGA-II, Deb et al. 2002), evolutionary hill climbing, and genetic programming.

As mentioned earlier, the Aldebaran Robotics Nao humanoid robot (Gouailler et al. 2009) is commonly used in recent research in this area, other platforms include the Robotis Bioloid humanoid robot, the iCub open-systems humanoid robot (Sandini et al. 2007), the Kondo KHR-2HV and the Fujitsu HOAP-1 and HOAP-2 robots.

The largest bipedal humanoid included in the survey is an adult-sized humanoid called Dexter, and developed by Anybots Inc. While successfully evolving walking in simulation on this robot, this walk was not successfully transferred to the real robot, however the evolved controller can balance the real robot in a standing position (Palmer et al. 2009).

To date the author is not aware of the successful application of evolutionary techniques directly to the design of complex motions in a sophisticated modern adult-size bipedal humanoid, such as the Cybernetic Human HRP-4C, described in Chap. 4. Further details of many of these humanoid robot platforms are also given in Chap. 4.

The most commonly used simulation platform is the Cyberbotics Webots mobile robot simulation package (Michel 2004). This allows for the simulation of a wide variety of robots and their environments, using an accurate physics simulator. Another commonly used simulator is the SimSpark simulator, as used in the

RoboCup 3D soccer simulation league (Obst and Rollman 2005; Boedecker and Asada 2008). Both the Webots and the SimSpark simulators use the Open Dynamics Engine (ODE) physics engine which allows for the simulation of complex rigid bodies connected with joints. The Urban Search and Rescue simulator (USARSim) has also been used to evolve behaviours (Carpin et al. 2007), as has OpenHRP (Open architecture humanoid robotics platform); (Kanehiro et al. 2004) and the Panda3D simulator (https://www.panda3d.org). A variety of custom simulators and mathematical models were also employed.

We present our findings in this section in tabular form for conciseness, and to allow for quick consultation and access by the interested reader to further relevant material, including the source publications. As for our discussion of research in the early years (2000–2007), we divide the experiments into two separate sections, representing simulated experimentation (no transference/experimentation on a real humanoid, Table 6.1), and embodied experimentation where a real robot was involved at some stage, either in evolution, or in the testing/implementation of evolved behaviours (Table 6.2). Each table also provides further details regarding the researchers involved in the work and the date of publication, in the case of embodied experimentation the humanoid robot platform (Nao, Bioloid, etc.), and in the case of both simulated and embodied experimentation the simulator employed (if specified). The general application area of the research is also outlined (bipedal locomotion, dance, etc.) along with the general type of fitness function(s) employed. The type of evolutionary algorithm is also given (GA, GP, etc.), and an indication is given as to what it is that the evolutionary algorithm actually evolves (weights for a neural network, motor rotation angles, etc.). It is hoped that this will provide sufficient information for readers to determine if a particular experiment is of interest to them, and to then provide them with the pointers to the relevant research literature, if required.

To recap, this section covers publications over the 6 years 2008–2013. It is not suggested that this is in any way a comprehensive listing of all of the publications during this period; however, it is hoped that this list forms a representative sample of applications, and gives the reader an overview of most of the important aspects of the current state of the art in the field of evolutionary humanoid robotics.

Chapter 7
Performance Evaluation and Benchmarking of Humanoid Robots

7.1 Introduction

In the evolutionary robotics field, or indeed in the general sphere of research into producing autonomous agents for real-world applications, it is desirable (if not essential) to have a situation where we can build on previous results in order to build/evolve artefacts of ever increasing utility. However, as pointed out in Eaton et al. (2001), if we are to produce progressively more useful robots, it is essential to have some method of comparing or benchmarking different performances. This is a difficult task given the wide variety of wheeled and legged mobile robot architectures and the diversity of the different tasks being attempted. In 2001, recognising the utility of aspects of the RoboCup competition model in the performance evaluation and benchmarking of mobile robots (together with RoboCup's inherent limitations), we then advocated (Eaton et al. 2001)

> the provision of a set of specifically designed experimental frameworks, based loosely on the RoboCup model, involving tasks of increasing complexity, rigorously defined to facilitate experimental reproducibility and verification

Del Pobil also discusses the issues involved in the performance evaluation and benchmarking of robots well in the rationale behind the "Benchmarks in Robotics Workshop", held at IROS (2006) (del Pobil 2006):

> Current practice of publishing research results in robotics makes it extremely difficult not only to compare results of different approaches, but also to assess the quality of the research presented by the authors …Typically when researchers claim that their particular algorithm or system is capable of achieving some performance, those claims are intrinsically unverifiable, either because it is their unique system or just because of a lack of experimental details, including a working hypothesis …Results are tested by solving a limited set of

© The Author(s) 2015
M. Eaton, *Evolutionary Humanoid Robotics*,
SpringerBriefs in Intelligent Systems, DOI 10.1007/978-3-662-44599-0_7

specific examples on different types of scenarios, using different underlying software libraries, incompatible problem representations, and implemented by different people using different hardware, including computers, sensors, arms, grippers ...

I was a member of the expert panel for Robot Standards and Reference Architectures (RoSta) FP6 EU Coordination Action, and one of its findings in 2007 was that (Hägele 2007)

> There are benchmarks for components and subsystems of mobile manipulation and service robots and also competitions in different areas but there is nothing like a whole system benchmark that produces comparable results

and

> There is a tremendous need for clearly defined benchmarks for systems with and without human interaction.

In this chapter we examine some of the advances which have been made in recent years in the performance evaluation and benchmarking of robots, concentrating mainly on those with some application in the humanoid robotics field. For a good overview of robotics challenges prior to 2002 see Balch and Yanco (2002).

The most recent, and the most ambitious initiative, which is currently under way, is the Defense Advanced Research Projects Agency (DARPA) Robotics Challenge. This follows on from the 2004 and 2005 DARPA Grand Challenges, and the 2007 DARPA Urban Challenge. It involves the robot (most likely, although not necessarily, of the humanoid variety) performing a range of different tasks including driving a utility vehicle, avoiding obstacles, manipulating objects, and opening a door.

We also look at the current RoboCup competition, concentrating on the Standard Platform League (SPL), which uses the Aldebaran Robotics Nao 21-DOF humanoid robot as described in Chap. 4 and the 3D simulation league, which uses a simulated model of the same Nao robot. Also of interest is the soccer humanoid league, which generally involves self-constructed humanoid robots in three different sizes: KidSize or less than 60 cm tall, TeenSize for robots which are between 100 and 120 cm, and the full AdultSize for robots taller than 130 cm. Also falling under the RoboCup umbrella are the more recent RoboCup@home, RoboCup@work, and RoboCupRescue competitions.

We also briefly discuss the FIRA HuroCup, which is aimed specifically at humanoid robots and which involves a variety of tasks including ball throwing, wall climbing, and penalty kicking. Other benchmarks of note, though not necessarily involving humanoid robots, include the ICRA Robot Challenge, the European Land Robot Trial (ELROB) (Schneider and Wildermuth 2011; Schneider et al. 2012), and the 2008 Rat's Life benchmark (Michel et al. 2008).

7.2 The DARPA Robotics Challenge

The DARPA Robotics challenge, introduced above, involves individual robots from different teams around the world competing in a variety of tasks based on an urban search-and-rescue environment. Parallels can be drawn here between the RoboCupRescue competitions, which drew inspiration for their creation from the 1995 Hanshi-Awaji earthquake, and the 2011 Great East Japan Earthquake, which caused the Fukushima nuclear power plant crisis, both of which could be seen as providing an impetus for the DARPA Robotics Challenge (DRC). Both of these situations were characterised by the need for robotic assistants to operate in difficult and dangerous (for humans) environments. The DARPA Robotics Challenge is of particular interest as, because of the nature of many of the tasks posed by the challenge, there may be an advantage in the robots having a humanoid or near humanoid body shape, although this is not specified in the rule set for the DRC.

The DRC comprises three sequential events: the Virtual Robotics Challenge (VRC), which took place in June 2013, the DRC trials, which took place in December 2013, and the DRC finals, which were due to take place originally in December 2014, but are now scheduled for June 2015. As part of the DRC, DARPA has funded the development of the Open Source Robotics Foundation (OSRF) Gazebo simulator together with the building of six Atlas robots (based on Boston Dynamics' PETMAN robot), which were donated to the teams that showed the best performance in the VRC.

7.2.1 The DRC Trials

For the DRC trials eight separate task sets were given, each generally divided into three separate subtasks, worth one point each. An extra bonus point is awarded for each task if the robot completes all of a task's subtasks without human intervention (apart from teleoperation). The robots were allowed 30 min to complete each given task. The time taken to complete tasks was not taken into account except in the case of a dead heat. The eight tasks to be performed (in summary) were:

7.2.1.1 The Vehicle Task

The robot has to drive a utility vehicle through an "obstacle course" consisting of a series of pylons (1 point), and then get out of the vehicle and walk/locomote to a defined "end zone" (2 points).

7.2.1.2 The Terrain Task

The robot must traverse three separate terrain segments consisting of blocks laid out on the ground (but not fastened to the ground). The successful transversal of each terrain segment earns the robot one point.

7.2.1.3 The Ladder Task

The robot climbs a ladder which is securely fastened to the floor, and set at either a 60° or a 75° angle, and which may or may not have handrails depending on team choice. Depending on the height climbed, the robot can get up to three points in total. This task posed particular difficulty for robots of non-humanoid morphology.

7.2.1.4 The Debris Task

The robot has to remove two groups of five pieces of debris consisting of light-weight wooden pieces; each group cleared gains one point. The robot must then travel through an open doorway (one point).

7.2.1.5 The Door Task

The robot must make its way through three separate doors, one point for each door. The doors are a push door, a pull door, and a door with a weighted closer. Each door is 36 inches wide, and has a lever-type handle.

7.2.1.6 The Wall Task

The robot has to operate a cordless drill in order to remove a triangular section from a half-inch thick sheet of drywall. One point is given for each of the first two edges cut through; the third point is for cutting the final edge and removing the triangular piece from the wall. The robot has a choice of two drills and two drill bits.

7.2.1.7 The Valve Task

This task involves the robot closing three different valve types, in any order. One valve is a 90° ball valve requiring a rotation of 90° to close fully. The other two valves are rotary valves of different diameters, each requiring a complete clockwise rotation to close. The successful closure of each valve garners one point.

7.2.1.8 The Hose Task

This task involves connecting a hose to a wye, again this is divided into three subtasks; moving the hose nozzle past a certain point, making contact between the hose nozzle and the wye, and attaching the hose nozzle to the wye. As usual, each of these subtasks is worth one point each.

7.2.2 The DRC Finals

Following the DRC trials, five of the top eight teams used Atlas-based robots and these automatically qualified for DARPA funding. The other three robots finishing in the top eight were the S-One humanoid from SCHAFT Inc., a humanoid broadly based on the HRP-2 humanoid, which came first in the trials, the Carnegie Mellon University robot CHIMP (a quadrupedal robot which is capable of bipedal standing and can roll on rubberised tracks and has "near-human form factor, work-envelope, strength and dexterity"), and the NASA Jet Propulsion Laboratory robot RoboSimian, probably the least human-like of the eight robots, which has four general-purpose limbs for both locomotion and manipulation.

In recent developments the top-scoring SCHAFT humanoid, which is now owned by Google, has been pulled out of the competition altogether (having earlier been moved to the self-funded category), and the funding freed distributed to two of the lower scoring teams one of which has split into two teams. It should be noted that while ten teams have been now been selected to receive DARPA funding for the 2015 DRC finals, any team can compete in these finals as long as they have independent funding. Team KAIST, from South Korea using the DRC-HUBO humanoid have confirmed their intention to compete in this self-funded category.

The tasks faced by the robots in the DRC Finals will be similar to those faced in the trials, with the significant difference that the robots will not be allowed to have any physical connections to their environment, be they restraints to prevent the robot from damage in the case of a fall, power connections, or wired communications. In addition no human intervention of any type will be allowed, for example, if a robot falls, the robot will have to get up without human assistance. As in the DRC trials the robots will operate on a teleoperated basis, however the wireless communications will be further degraded in order to simulate the conditions of wireless interference and latency that might be experienced in a real disaster zone. Finally, the time allowed to complete the tasks will be reduced to about a quarter of that allowed for the DRC trials.

7.3 RoboCup

Following several years of preparatory work, the RoboCup soccer league was launched in Nagoya, Japan in 1997 with the avowed aim (Kitano and Asada 1998; Kitano et al. 1998)

> By mid-21st century, a team of fully autonomous humanoid robot soccer players shall win the soccer game, comply with the official rule of the FIFA, against the winner of the most recent World Cup

This goal was viewed as a grand challenge and would certainly have seemed ambitious at the time. However, the given timescale of around 50 years was viewed in the context of the approximately 50 year timescale between the invention of the digital computer and the building of a computer which could beat the human world champion in chess. The computer, of course, was Deep Blue, and the champion Garry Kasparov; and this landmark occurred in 1997, the year the RoboCup initiative was launched.

While the initial robots participating in the RoboCup competition were all of the wheeled variety (or simulated), the intention was always to progress to humanoid robots, which would then lead to the ultimate goal of the challenge: that is, to take on and to beat the human World Cup champions (Kitano and Asada 1998). Following this longstanding vision, the first demonstration of humanoid robot soccer skills took place in the 2000 RoboCup competitions, which was the followed by the first RoboCup Humanoid League in 2002 (Veloso and Stone 2012).

The current version of the RoboCup Humanoid league is divided into three sections: the KidSize (40–90 cm tall), the TeenSize (80–140 cm) and the AdultSize (130–180 cm). The KidSize class currently consists of two teams of four humanoid robots each competing against each other; the TeenSize has two teams of two humanoids each, while the AdultSize league (only in operation since 2010) currently involves two adult-size humanoids competing against each other in dribble and kick-type competitions. Additional technical challenges are also included as part of the humanoid league, including self-localisation, stable walking and running, ball perception, and ball kicking. The design of the robots for all of the classes is constrained from a morphological perspective to having a human-like body with the only allowed modes of locomotion being bipedal walking and running. Similarly, the design of sensors for the robots is constrained to be broadly human-like; for example, any cameras used must be positioned in the robot's head and the field of the robot's view is limited to 180°.

In order to fulfill the vision of having a fully autonomous humanoid team capable of playing (and beating!) the human world champions by the year 2050, a tentative roadmap has recently been proposed, setting out a series of rough milestones that would have to be achieved in order to achieve this goal. This roadmap involves the gradual increase of the number of players per team, the size of the playing field, the number of humanoid robot players, and the length of each individual contest. It is also proposed to reduce the number of classes for all sizes of robots to just one. So by the year 2020, for example, it is proposed that there will be a single class of humanoid, with a minimum height of 60 cm (about the size of the current Nao humanoid) with a maximum of 6 players per side (currently the maximum is 4) and double the time per half (20 min as opposed to 10 min). By the year 2030 the minimum robot height will have risen to a metre with a field length of 80 m and 8 players per side. It is also envisioned at this stage that the robot players will engage in competitions against (nonprofessional!) human teams. In addition, at this stage an extra technical challenge as well as defending the goal and striking at goal against human opponents will be to be able to outrun the president of RoboCup!

While the Humanoid league involves the (human) competitors generally designing and building their own humanoid robots, the current RoboCup Standard Platform League (SPL), while involving humanoid robots (since replacing the discontinued Sony Aibo quadrupedal robot in 2008), specifies that these robots share a common hardware platform. All of the teams in the SPL use Aldebaran Nao humanoid robots, so the emphasis in the SPL is on software competition (Barrett et al. 2013). As a result, there has been a lot of experimentation in recent years in the application of evolutionary (and other) techniques to the optimisation of the soccer-playing capabilities of Nao robots.

As a testament to the growing popularity of the event, RoboCup 2013, held in the Netherlands, attracted some 40,000 visitors, and had over 2,500 participants from 45 countries participating in the various leagues and in the associated RoboCup symposium. Interestingly, of the major leagues, the humanoid league (excluding the Nao based SPL) had both the highest number of participants (285) and the largest number of teams (38).

7.3.1 RoboCup@Home/RoboCup@Work

Associated with the RoboCup soccer league are the RoboCup@Home league, and the recently established RoboCup@Work league. The RoboCup@Home league was established in 2006 in order to encourage the development, and to aid in the benchmarking of robots that will be able to function in typical household environments. While these robots are not constrained to be of humanoid form, because they are designed to operate in environments humans find themselves in, their overall size and typically upper body structure tend to be broadly humanlike. Typical skills tested for include safe locomotion around indoor environments, object recognition, grasping, and human-robot-interaction (HRI). Robots have also been tested in recent years for their ability to operate outside the home environment, for example, as shopping assistants to humans in real store environments (Stückler et al. 2013). The associated RoboCup@Work League aims at the development of autonomous and flexible robots for safe operation in industrial rather than in domestic environments, for example, in the cooperative collection and/or transportation of objects (with human and/or other robot assistance), the operation of machinery, and the loading/unloading of containers (Leibold et al. 2013).

7.3.2 RoboCupRescue

Inspired by the devastating 1995 Hanshi-Awaji earthquake, which caused great damage and loss of life in Kobe City, Japan, the RoboCupRescue Robot League was founded in order to Röfer et al. (2012); preface

> promote the development of intelligent, highly mobile, dexterous robots that can improve the safety and effectiveness of emergency responders performing hazardous operational tasks.

In RoboCupRescue, teams of robots search for signs of life in a variety of colour-coded terrains of varying levels of difficulty using robots which may operate on an autonomous basis, under human operator control, or a combination of these two modes. A recommended maximum height of 0.7 m is considered desirable in order to negotiate tight terrains but this is not enforced. Some of the more difficult challenges posed include crossing over ramps and negotiating stairs and a steep inclined plane. There is also an associated Rescue Simulation League.

7.4 FIRA HuroCup

The Federation of International Robot-soccer Association (FIRA) was established at an international robot soccer tournament in KAIST, Daejeon, Korea in 1997. While FIRA-organised international robot soccer competitions are still regularly held, RoboCup has grown over the years to be the largest and most prestigious competition. Of more interest to us is the associated FIRA HuroCup competition, which is the longest running intelligent humanoid robot competition, with the initial contest involving five teams taking place in 2002 (Anderson et al. 2011). Originally designed as a robot soccer competition for humanoid robots, the focus quickly changed to address those open research problems particularly associated with humanoid robots. The FIRA HuroCup robotics challenge currently consists of an octathlon of separate events including a sprint (3 m distance) and a marathon (about 42 m distance), penalty kick, weight lifting, a basketball challenge (pick up a ball and throw in a basket) and the recently introduced wall climbing challenge. All robots must operate in fully autonomous mode, and the variety of tasks involved (all of which must be performed by a single robot) reduces the likelihood of the use of highly specialised hardware that may cause the robot to perform better in one individual task, to the potential detriment of other task(s). For example, a robot that employs specialised hardware in order to compete more effectively in the basketball event may perform more poorly in, for example, the sprint event, because of the additional weight carried by the robot (Anderson et al. 2011).

7.5 Summary and Conclusions

In this chapter we have addressed the thorny problem of how to evaluate and benchmark the performance of robots which, although perhaps humanoid in their gross morphology, may be very different in many other ways. We have considered the importance of some form of benchmarking process in order to evaluate different software and/or hardware architectures, whether arrived at by an evolutionary process, or by some other means. Over the years, the RoboCup initiative particularly has provided the impetus for the development of a range of skill sets, albeit in a rather artificial and constrained environment. RoboCup@home and RoboCupRescue attempt to bring robots out into more real-world environments.

Another interesting recent initiative is the HUMABOT challenge, launched at the 2014 IEEE International Conference on Humanoid Robots, which is aimed specifically at humanoid robots of similar size to the Nao robot, and is based in the (scaled) kitchen of a house. This new challenge poses tasks such as climbing stairs, simple meal preparation, and creating a shopping list through recognising which products are present, and which are absent, from the kitchen shelves. There are also several ongoing European projects specifically aimed at the investigation, creation, evaluation, and benchmarking of human-like motions (Torricelli et al. 2014).

Certainly with the DARPA Robotics Challenge there is a sense of history in the making: robots mainly, though not exclusively, humanoid, performing a variety of tasks which only a decade or so ago would have been considered in the realms of science fiction.

And this may, indeed, only be the beginning. In the words of the DARPA program manager Gill Pratt

> I think part of the good that can come out of the [DARPA Robotics] trials is that we'll actually help calibrate the public to what the reality is in this field. Part of the difficulty with science fiction is that if there's no counterexample of science fact, people can get the idea that these [robots] aren't very hard to build. So, besides calibrating ourselves to what the state of the art is, I think a lot of the good that we can do here is to calibrate the public.

Which brings us along nicely to the subject matter of the next chapter—what exactly should we expect, in the near and not-so-distant future, from our evolving humanoid companions? Is it reasonable to expect that the general public, while being "calibrated" by the achievements of highly talented engineers in advanced robotics, should perhaps also have some input as to the shape and the applications of this technology, given its potentially profound and far-reaching consequences for humanity as a whole?

Chapter 8
Ethical, Philosophical and Moral Considerations

8.1 Introduction

It may seem a little strange in a book on advanced robotics and control systems to have a section on such "soft" issues as ethics, philosophy, and morality. I would, however, make the contrary argument: might it be strange not to discuss, at least in outline, the potential ramifications for human society of the creation, by an evolutionary process, of mechanical creatures in our own likeness whose workings may not (and may never) be fully understood, a fact which some of the earliest researchers in this field readily acknowledged (de Garis 1990a; Sims 1994a).

So, given the potential far-reaching consequences of research in the field of humanoid robotics, and in evolutionary humanoid robotics in particular, where we may evolve humanoid robot architectures of advanced capabilities, perhaps even approaching human-level abilities in some aspects, this author considers it to be appropriate, if not essential, to have at least a cursory discussion of these issues. Whatever unease may be felt by some readers at its inclusion may be alleviated somewhat by the relative brevity of this chapter.

Since the publication of Charles Darwin's *On the Origin of Species* (Darwin 1859) writers have been discomfited by the thought of the potential evolution of machinery which could be the intellectual equivalent of humans, or indeed their superiors. Samuel Butler in his classic novel *Erewhon, or Over the Range*, originally published in 1872 not very long after Darwin's monumental work, described a fictional world which was devoid of machines because of the fear that "machines were ultimately destined to supplant the race of man", by developing their own consciousness by means of Darwinian selection (Butler 1970).

In more recent times, one of the pioneering researchers in the field of evolutionary humanoid robotics, Hugo de Garis, who conducted some of the earliest

© The Author(s) 2015
M. Eaton, *Evolutionary Humanoid Robotics*,
SpringerBriefs in Intelligent Systems, DOI 10.1007/978-3-662-44599-0_8

experiments in the application of evolutionary techniques to the problem of bipedal locomotion, has predicted a major future global war before the end of this century between two factions, those who wish to construct hyperintelligent creatures, or "artilects", and those who fear that these creatures will seek world domination, with potentially dire consequences for humanity (de Garis 2005).

8.2 Man Versus Machine

By all means celebrate and encourage traits such as punctuality and conformity to societal norms in situations where this is required to preserve an orderly society. In equal measure, however, we should also celebrate our essential humanity—that trait which binds together, to a greater or lesser extent, the seven billion or so human inhabitants of this small planet. This celebration of our essential humanity will become more and more important, as we are now approaching the stage where we may, in the relatively near future (years, not decades), have humanoids very closely visually representing humans, but with potentially far superior physical, and in certain fields, intellectual capabilities.

As we have already mentioned, the researcher Hugo de Garis touts the possibility—indeed, the near certainty—of a future global "gigawar" (billions of dead) resulting from a conflict between those of us who wish to construct "godlike" robots and those of us who don't, seeing this development as leading to the near-inevitable demise of humanity as we know it. While ideas such as those of de Garis may be seen as on the extreme side of the academic spectrum of thought on this issue, there is no doubt that there are many researchers with mixed ideas about the potential advantages of future robotic technologies. We must remember that evolution is not, in itself, a benign force—it has resulted in warlike humankind. Do we now wish to be unwitting authors of our own destruction? This is an idea that is a little farfetched maybe, but perhaps worthy of some consideration.

As the futurist and inventor Ray Kurzweil puts it, as quoted in James Barrat's recent thought-provoking text *Our Final Invention: Artificial Intelligence and the End of the Human Era* (Barrat 2013, p. 265)

> Machines will follow a path that mirrors the evolution of humans. Ultimately, however, self-aware, self-improving machines will evolve beyond humans' ability to control or even understand them.

At present, this is truly the stuff of science fiction, but some areas of science fiction have a rather unsettling tendency to become science fact, and in many cases in a far shorter time span than the science fiction writer may have predicted. At least armed with some idea of the possible future potential of these technologies, we can now begin to take steps to try to ensure that these technologies will be used for the betterment, and not for the long-term detriment, of mankind.

8.3 The Future Really Doesn't Need Us?

In the year 2000 Bill Joy, then Chief Scientist at Sun Microsystems, wrote a widely quoted article for *Wired* magazine entitled "Why the Future Doesn't Need Us" (Joy 2000), in which he contends (in the subtitle of this article) that

> Our most powerful 21st-century technologies—robotics, genetic engineering, and nanotech—are threatening to make humans an endangered species.

While in the course of the preparatory work for the writing of this book, the author came across a "blog" (web log) relating to the developing humanoid robot research field suggesting a sinister underlying agenda. This went along the following lines:

- Make humans think we are "humanoids" by developing ever more complex and capable humanoid robots.
- Replace many awkward and questioning humans with humanoids.
- The state/country/world now becomes much easier to govern by the "elite class" of (presumably) humans.

These observations, together with the unquestionable fact that a future potential application of "intelligent" humanoid robots, such as those which make use of a process of artificial evolution as part of the design process of body and/or brain, will undoubtedly be of a military nature, did make this author question the propriety of writing a text of this nature in the first place. However whatever misgivings I may have on this front, the technologies involved will march ahead in any case, so, as in many areas of technological advancement such as nanotechnology and the biosciences, it is up to humanity itself to decide whether to use for good or for ill.

8.4 Fear of the Machine?

It seems to be quite fashionable nowadays to be almost embarrassed about being human, and to having a humanoid shape. Machines are so prevalent and so powerful, both on an "intellectual" level (computers, smart phones, etc.), and in a physical sense (modern automated cars, factory robots, etc.); and of course there is also the notion that we have arrived at our current body shape by, in a certain sense, a series of evolutionary "accidents".

Much of this may be true, but what certainly is true is that we do (or our "consciousness" does) currently inhabit a humanoid shape which is consistent among the many different sentient races of humanity. There is also a clear avenue of thinking that says that the intellectual formation of creatures, both natural and artificial, is, in a large part, determined by the body shape they inhabit, as clearly elucidated by Pfeifer and colleagues' aptly named text *How the Body Shapes the Way We Think* (Pfeifer et al. 2007).

So if we, for whatever reason, wish to recreate intelligence that will operate in a humanlike fashion, it makes sense to generate this intelligence for a humanoid robot, for maximum effectiveness. And, taken to its extreme, this book is about the potential of using a process based on natural evolution to create artificial analogues of ourselves, no more, no less. On this level we might characterise a situation where we create creatures, in our own likeness, with intellectual and physical abilities similar to, or even exceeding, our own, through a process which, on one level, we do not fully understand.

Should we be afraid? Certainly. But at least with the fear that knowledge brings we can take action, if this is deemed necessary. Without this knowledge, presently freely available to those who choose to acquire it, the future for us, humanity, is far more bleak.

There are conceivably those who, for their own reasons, might wish to minimise the exposure of the general public to the startling advancements currently being made in humanoid robots, and in associated areas. It is not the function of this book to cause fear or anxiety to the general public. Its function is to inform, and it is, as such, particularly aimed at researchers in the AI and robotics fields who desire a concise introduction to this area, together with educators (myself included) who may find this book as a useful primary, or ancillary, text in their AI/robotics advanced undergraduate or graduate course(s).

However, there may well be concerns, among academics, practitioners, and among the general public alike, about the potential implications for the public good of humanoids of high intellect, whose functioning we do not fully comprehend.

Turing put it quite concisely in his 1948 essay, "Intelligent Machinery", as reproduced in Copeland (2004), where he envisaged a creature for which

> In order that (the machine) should have a chance of finding things out for itself it should be allowed to roam the countryside and the danger to the ordinary citizen would be serious … This method … seems to be altogether too slow and impracticable.

A little while after this, a compatriot of Turing's, the English mathematician I.J. Good, commented in his article "Speculations Concerning the First Ultraintelligent Machine" (Good 1965)

> The first ultraintelligent machine is the *last* invention that man need ever make, provided that the machine is docile enough to tell us how to keep it under control. It is curious that this point is made so seldom outside of science fiction. It is sometimes worthwhile to take science fiction seriously.

8.5 Technology Taking Over?

With the advent of modern technologies (advanced robotics, biotechnology, nanotechnology, etc.), all of which have the potential to fundamentally alter the face of humanity, perhaps it is time now to consider the importance of the possible regulation of these increasingly important and potentially invasive fields of research.

It could be argued that we, as humans, are being increasingly compartmentalised and constrained in both a spatial (for example, more high-density housing in the large conurbations) and in a temporal sense (being expected to conform to rigid timescales and deadlines which are not in tune with natural bodily rhythms). In addition there is, of course, the near-constant analysis of our movements (ubiquitous CCTV cameras, especially in urban centers), our purchasing habits (so-called loyalty cards, etc.), and of course online where every activity can be potentially monitored and logged. Much of this surveillance is, of course, aimed at categorising and pigeonholing people as consumers, for marketing purposes. As a byproduct, there is more and more subtle pressure to conform, to behave in a stereotyped fashion appropriate to the particular pigeonhole in which one has been placed. Machines, of course, have no difficulty with such constraints. They are happy to operate 24/7 (subject to maintenance) in constrained conditions which no human would find bearable.

Indeed, some would argue that technology has now turned the corner, that from being a great enabling force for mankind, allowing the realisation of dreams and ambitions that, only a century or so ago, would have been considered by even the most far-thinking as being the imaginings of a delusionist rather than a visionary, to being in many instances a force that is self-driving, pushing humans into corners that they do not want to go. This, then, is one of the great challenges facing modern humankind, how to unleash the undoubted gigantic benefits of future enabling technologies, without falling into the ultimate trap—the very loss, no less, of humanity itself, as we know it.

8.6 A Way Forward?

However, there may be a way forward. Instead of viewing this burgeoning technology as a threat and as a potential future replacement for *homo sapiens*, let us use it to our advantage. This is already being done in many walks of life, with machines taking over back-breaking and tedious manual jobs. Let us see this new technology (up to and including the possible development in the future of humanoid robots approximating human levels of intelligence and dexterity), as an aid to human society in liberating us from the tedious chores and short lifespan of the past to a world where we can celebrate more fully our existence on this earth. This difference in emphasis is subtle, yet important. Ethical issues may well arise in the distant future as to the treatment of these future evolved (or otherwise designed) advanced humanoid robots; however, these issues should not prove insurmountable.

One simple initial measure to encourage the use of these revolutionary new technologies for good rather than for destructive purposes would be an outright ban on the use of future autonomous humanoid and other intelligent robots for military purposes. This would be akin to the current global bans on the use of chemical and biological weapons.

Of course, one may ask whether there is there really any difference between lethal killing machines in humanoid form and malevolent robots which look nothing like humans. This is, to a certain extent, a philosophical issue. What is most certainly not a philosophical issue is the question of the desirability of autonomous intelligent agents with lethal capability possessing humanlike mobility.

This situation is becoming more serious with predictions from some sources that within a very short time there will exist (tele-operated/semiautonomous) robots in humanoid form capable of operating alongside existing human soldiers in the battlefield. The next step is, of course, fully autonomous humanoid robots with lethal capabilities. The question arises: is this what we want, as humans?

Norbert Weiner, the father of cybernetics as a field, who we quoted in Chap. 2, had severe misgivings about the potential future ramifications of the field in general. In the introduction to his seminal text *Cybernetics: or Control and Communication in the Animal and the Machine* (Wiener 1948) he states his position clearly

> Those of us who have contributed to the new science of cybernetics ...have contributed to the initiation of a new science which... embraces technical developments with great possibilities for good and for evil ... We do not even have the choice of suppressing these new technical developments ... The best we can do is to see that a large public understands the trend and the bearing of the present work, and to confine our personal efforts to those fields ... most remote from war and exploitation.

In more recent times, one of the plenary talks given at the recent IEEE Humanoids 2013 conference, a flagship conference for those involved in humanoid robotics research, was given by the eminent researcher and author Ronald Arkin of Georgia Institute of Technology. The intriguing title of his talk was "How to NOT Build a Terminator", which referenced the ongoing DARPA Robotics Challenge discussed in the previous chapter, in which he argued that the "Terminator" robot of science-fiction lore is rapidly becoming a potential reality, with all that this implies.

Because of the potential future threats posed, the international Human Rights Watch (HRW) organisation and the International Human Rights Clinic (IHRC) at the Harvard Law School recommend, among other measures, that all states

- Prohibit the development, production, and use of fully autonomous weapons through an international legally binding instrument.
- Adopt national laws and policies to prohibit the development, production, and use of fully autonomous weapons.
- Commence reviews of technologies and components that could lead to fully autonomous weapons.

We are at an important stage in human history where decisions made now could have profound implications far into the future. Do we celebrate humanity, or else abandon ourselves to whatever new technology produces? As scientists and as engineers we have a responsibility to let our opinions clearly be known on these matters.

In the words of Francis Fukuyama, speaking clearly on the different but not unrelated topic of biotechnological advances (Fukuyama 2002)

> We do not have to regard ourselves as slaves to inevitable technological progress, when that progress does not serve human ends.

and

> True freedom means the freedom of the political community to protect the values they hold most dear …

Perhaps the time has come for us all, not just AI researchers, roboticists, and nano/biotechnologists, but humanity in general, to decide what these values are, and to take firm and decisive action to ensure their protection now, and for future generations.

Chapter 9
Conclusions, and Looking to the Future

In this book we have seen the application of evolutionary techniques to the design of both the "body" and the "brain" of humanoid robots, from simple stick-leg simulations in the early 1990s to current applications involving the implementation of increasingly complex behaviours on real humanoid robots with many degrees of freedom. It is probably fair to say that this is only the beginning of the story. As long as humanity survives in its present form, humanoid robots are here to stay. Whether as companions or as slaves, as entertainers or even as killers, they will be in our presence, perhaps in some cases even going unnoticed except to the trained eye. And, as we strive to imbue these robots with ever more versatile and humanlike qualities, evolutionary algorithms, inspired by the natural forces that conspired to produce us, their inventors, will very likely serve as an invaluable tool in this regard. However, there is no question that along the way resistance will be met to the notion of evolving creatures in our own likeness, whose functioning we do not fully understand, and with the associated ethical and societal issues briefly discussed in the previous chapter. However, undoubtedly, used wisely, this new technology could be of great benefit to humankind.

9.1 Beyond Humanoids—Where to Next?

If we are to accept our "loose" definition of a humanoid robot as a robot designed to be able to operate in conditions and environments that humans function in (level n: built-for human, BFH), we can include robots possessing high levels of dexterity, intelligence, and strength, albeit generally restricted to a limited task set. Most humanoid robots to date fall well short of humans in each of these categories, certainly in the areas of intelligence and dexterity. Specifically on the subject of dexterity, or the ability to manipulate the environment in complex ways, interacting with objects of different shapes and sizes, most robots today fall short of human-comparable capability. However it is possible to speculate (and this is certainly mere speculation at this stage) about future "Humanoid" robots of a highly dexterous nature, perhaps possessing multiple sets of limbs, more than one method

© The Author(s) 2015
M. Eaton, *Evolutionary Humanoid Robotics*,
SpringerBriefs in Intelligent Systems, DOI 10.1007/978-3-662-44599-0_9

of locomotion (wheeled, legged, etc.) with multiple sensory inputs, and with abilities far beyond the capabilities of humans, yet designed to operate in BFH environments.

Looking beyond the notion of "what is" to "what could be", one can envisage the notion of dynamically reconfigurable morphology/hardware; that is, the situation where the robot morphology would change dynamically as the evolution process progresses. Even now some progress is being made on the coevolution of body and brain of sophisticated robots by a number of researchers, although their efforts are not currently specifically targeted at humanoid robot design. In particular, recent research has indicated that in relation to grasping and object manipulation tasks in particular there are definite advantages to the coevolution of both the robot's morphology and control systems (Bongard 2010).

Another intriguing possibility for future generations of advanced humanoid robots arises where these robots, if equipped with suitably designed advanced manipulators (robotic arms/fingers, etc.) and an appropriately designed workshop, might well be able to alter their own detailed (or indeed gross) morphology in response to either immediate demands from the environment, or to long-term evolutionary pressure. Whether or not such exotic robots would find widespread acceptance is, of course, quite a different matter.

We may draw a certain analogy with the development of early computer systems (before the advent of ubiquitous computer-aided design techniques). The development of each new generation of computers was a process of slow and painstaking progress. Nowadays, with the advances in computer-aided design technologies computers are, in a very real sense, helping themselves to create the next generation of computing, which will in turn make the current generation of computers running these programs obsolete. Huge advances are being made that would be impossible without the bootstrapping aid of current technology. Similarly, one could argue, when humanoid robots advance to the point where they themselves can play a part in the development of future humanoid robot generations we may see a parallel explosion of advancements.

This, certainly, at this moment, is the stuff of science fiction, but to paraphrase I. J. Good—sometimes it is worthwhile to take science fiction seriously. Who knows what possibilities ever-accelerating advances in technology may afford, for good or for evil? It is up to us all, scientists, engineers, and the public at large, to decide.

References

Akhtaruzzaman, M., & Shafie, A. A. (2010). Advancement of android and contribution of various countries in the research and development of the humanoid platform. *International Journal of Robotics and Automation (IJRA), 1*(2), 43–57.

Al Borno, M., De Lasa, M., & Hertzmann, A. (2013). Trajectory optimization for full-body movements with complex contacts. *IEEE Transactions on Visualization and Computer Graphics, 19*(8), 1405–1414.

Alankus, G., Bayazit, A. A., & Bayazit, O. B. (2005). Automated motion synthesis for dancing characters. *Computer Animation and Virtual Worlds, 16*(3–4), 259–271.

Allen, B. F., & Faloutsos, P. (2009a). Evolved controllers for simulated locomotion. In *Motion in games* (pp. 219–230). Berlin, Heidelberg: Springer.

Allen, B., & Faloutsos, P. (2009b). Complex networks of simple neurons for bipedal locomotion. In *IEEE/RSJ International Conference on Intelligent Robots and Systems, 2009. IROS 2009* (pp. 4457–4462).

Amor, H. B., Berger, E., Vogt, D., & Jung, B. (2009). Kinesthetic bootstrapping: Teaching motor skills to humanoid robots through physical interaction. In *KI 2009: Advances in Artificial Intelligence* (pp. 492–499). Berlin, Heidelberg: Springer.

Anderson, J., Baltes, J., & Cheng, C. T. (2011). Robotics competitions as benchmarks for AI research. *The Knowledge Engineering Review, 26*(01), 11–17.

Antonelli, M., Dalla Libera, F., Menegatti, E., Minato, T., & Ishiguro, H. (2009). Intuitive humanoid motion generation joining user-defined key-frames and automatic learning. In *RoboCup 2008: Robot soccer world cup XII* (pp. 13–24). Berlin, Heidelberg: Springer.

Arakawa, T., & Fukuda, T. (1996). Natural motion trajectory generation of biped locomotion robot using genetic algorithm through energy optimization. In *IEEE International Conference on Systems, Man, and Cybernetics, 1996* (Vol. 2, pp. 1495–1500).

Arakawa, T., & Fukuda, T. (1997). Natural motion generation of biped locomotion robot using hierarchical trajectory generation method consisting of GA, EP layers. In *Proceedings of the 1997 IEEE International Conference on Robotics and Automation 1997* (Vol. 1, pp. 211–216).

Azarbadegan, A., Broz, F., & Nehaniv, C. L. (2011). Evolving Sims's creatures for bipedal gait. In *IEEE Symposium on Artificial Life (ALIFE), 2011* (pp. 218–224).

Balakirsky, S., & Kootbally, Z. (2012). USARSim/ROS: A combined framework for robotic control and simulation. In *ASME/ISCIE 2012 International Symposium on Flexible Automation, American Society of Mechanical Engineers* (pp. 101–108).

Balch, T., & Yanco, H. (2002). Ten years of the AAAI mobile robot competition and exhibition. *AI Magazine, 23*(1), 13.

Bar-Cohen, Y., & Hanson, D. (2009). *The coming robot revolution: Expectations and fears about emerging intelligent, humanlike machines*. New York: Springer.

Barrat, J. (2013). *Our final invention: Artificial intelligence and the end of the human era*. London: Macmillan.

© The Author(s) 2015

M. Eaton, *Evolutionary Humanoid Robotics*,

SpringerBriefs in Intelligent Systems, DOI 10.1007/978-3-662-44599-0

Barrett, S., Genter, K., He, Y., Hester, T., Khandelwal, P., Menashe, J., et al. (2013). UT Austin
   Villa 2012: Standard platform league world champions. In *RoboCup 2012: Robot soccer world
   cup XVI* (pp. 36–47). Berlin, Heidelberg: Springer.

Baydin, A. G. (2008). Evolution of central pattern generators for the control of a five-link planar
   bipedal walking mechanism (No. arXiv: 0801.0830).

Baydin, A. G. (2012). Evolution of central pattern generators for the control of a five-link bipedal
   walking mechanism. *Paladyn, 3*(1), 45–53.

Beer, R. D., Chiel, H. J., & Sterling, L. S. (1990). A biological perspective on autonomous agent
   design. *Robotics and Autonomous Systems, 6*(1), 169–186.

Beer, R. D., & Gallagher, J. C. (1992). Evolving dynamical neural networks for adaptive behavior.
   Adaptive Behavior, 1(1), 91–122.

Berger, E., Amor, H. B., Vogt, D., & Jung, B. (2008). Towards a simulator for imitation learning
   with kinesthetic bootstrapping. In *Workshop Proceedings of International Conference on
   Simulation, Modeling and Programming for Autonomous Robots (SIMPAR)* (pp. 167–173).

Beyer, H. G., & Schwefel, H. P. (2002). Evolution strategies—A comprehensive introduction.
   Natural Computing, 1(1), 3–52.

Boedecker, J., & Asada, M. (2008). Simspark—concepts and application in the RoboCup 3D
   soccer simulation league. In *Proceedings of SIMPAR-2008 Workshop on the Universe of
   RoboCup Simulators, Venice, Italy* (pp. 174–181).

Boeing, A. (2008). Morphology independent dynamic locomotion control for virtual characters. In
   IEEE Symposium on Computational Intelligence and Games, 2008. CIG'08 (pp. 283–289).

Boeing, A., & Bräunl, T. (2002). Evolving splines: An alternative locomotion controller for a
   bipedal robot. In *7th International Conference on Control, Automation, Robotics and Vision,
   2002. (ICARCV 2002)* (Vol. 2, pp. 798–802).

Boeing, A., & Bräunl, T. (2007). Evaluation of real-time physics simulation systems. In
   *Proceedings of the 5th International Conference on Computer Graphics and Interactive
   Techniques in Australia and Southeast Asia* (pp. 281–288). ACM.

Boeing, A., & Bräunl, T. (2012). Leveraging multiple simulators for crossing the reality gap. In
   12th International Conference on Control Automation Robotics and Vision (ICARCV), 2012
   (pp. 1113–1119).

Boeing, A., Hanham, S., & Bräunl, T. (2004). Evolving autonomous biped control from simulation
   to reality. In *Proceedings of the 2nd International Conference on Autonomous Robots and
   Agents, Palmerston North, New Zealand* (pp. 13–15).

Bongard, J. C. (2010). The utility of evolving simulated robot morphology increases with task
   complexity for object manipulation. *Artificial Life, 16*(3), 201–223.

Bongard, J. C. (2013). Evolutionary robotics. *Communications of the ACM, 56*(8), 74–83.

Bongard, J. C., & Lipson, H. (2004). Once more unto the breach: Co-evolving a robot and its
   simulator. In *Proceedings of the Ninth International Conference on the Simulation and
   Synthesis of Living Systems (ALIFE9)* (pp. 57–62).

Bongard, J. C., & Paul, C. (2001). Making evolution an offer it can't refuse: Morphology and the
   extradimensional bypass. In *Advances in artificial life* (pp. 401–412). Berlin, Heidelberg:
   Springer.

Brooks, R. A. (1991). New approaches to robotics. *Science, 253*(5025), 1227–1232.

Brooks, R. A. (1992). Artifical life and real robots. In *Toward a Practice of Autonomous Systems:
   Proceedings of the 1st European Conference on Artificial Life* (pp. 3–10).

Brooks, R. A., Breazeal, C., Marjanović, M., Scassellati, B., & Williamson, M. M. (1999). The
   Cog project: Building a humanoid robot. In *Computation for metaphors, analogy, and agents*
   (pp. 52–87). Berlin, Heidelberg: Springer.

Brooks, R. A., Aryananda, L., Edsinger, A., Fitzpatrick, P., Kemp, C., O'Reilly, U.-M., et al.
   (2004). Sensing and manipulating built-for-human environments. *International Journal of
   Humanoid Robotics, 1*(01), 1–28.

Butler, S. (1970). Erewhon, or: Over the Range [1872]. *Harmondsworth: Penguin, 72*, 74–75.

Capi, G., Nasu, Y., Barolli, L., Mitobe, K., & Takeda, K. (2001a). Application of genetic algorithms for biped robot gait synthesis optimization during walking and going up-stairs. *Advanced Robotics, 15*(6), 675–694.

Capi, G., Nasu, Y., Barolli, L., Mitobe, K., & Yamano, M. (2001b). Real time generation of humanoid robot optimal gait for going upstairs using intelligent algorithms. *Industrial Robot: An International Journal, 28*(6), 489–497.

Capi, G., Yokota, M., & Mitobe, K. (2005). A new humanoid robot gait generation based on multiobjective optimization. In *Proceedings of the 2005 IEEE/ASME International Conference on Advanced Intelligent Mechatronics* (pp. 450–454).

Capi, G., Yokota, M., & Mitobe, K. (2006). Optimal multi-criteria humanoid robot gait synthesis—an evolutionary approach. *International Journal of Innovative Computing, Information and Control, 2*(6), 1249–1258.

Cardenas-Maciel, S. L., Castillo, O., & Aguilar, L. T. (2011). Generation of walking periodic motions for a biped robot via genetic algorithms. *Applied Soft Computing, 11*(8), 5306–5314.

Cariani, P. (1989). On the design of devices with emergent semantic functions (Doctoral dissertation, State University of New York).

Carpin, S., Lewis, M., Wang, J., Balarkirsky, S., & Scrapper, C. (2007). USARSim: A robot simulator for research and education. In *Proceedings of the IEEE International Conference on Robotics and Automation ICRA* (pp. 1400–1405).

Cecconi, F., & Parisi, D. (1989). Networks that learn to predict where the food is and also to eat it Proc. IJCNN (Washington, DC) vol 2 (Piscataway, NJ, IEEE) p. 624.

Chen, L., Yang, P., Liu, Z., Chen, H., & Guo, X. (2007). Gait optimization of biped robot based on mix-encoding genetic algorithm. In *2nd IEEE Conference on Industrial Electronics and Applications, 2007. ICIEA 2007* (pp. 1623–1626).

Cheng, M. Y., & Lin, C. S. (1995). Genetic algorithm for control design of biped locomotion. In *IEEE International Conference on Systems, Man and Cybernetics, 1995. Intelligent Systems for the 21st Century* (Vol. 2, pp. 1315–1320).

Cheng, M. Y., & Lin, C. S. (1997). Genetic algorithm for control design of biped locomotion. *Journal of Robotic Systems, 14*(5), 365–373.

Choi, S. H., Choi, Y. H., & Kim, J. G. (1999). Optimal walking trajectory generation for a biped robot using genetic algorithm. In *Proceedings of the IEEE/RSJ International Conference on Intelligent Robots and Systems 1999. IROS '99* (Vol. 3, pp. 1456–1461).

Cisneros, R., Yoshida, E., & Yokoi, K. (2012). Ball dynamics simulation on OpenHRP3. In *IEEE International Conference on Robotics and Biomimetics (ROBIO), 2012* (pp. 871–877).

Copeland, B. J. (Ed.). (2004). *The essential Turing: Seminal writings in computing, logic, philosophy, artificial intelligence, and artificial life, plus the secrets of Enigma.* Oxford: Clarendon Press.

Darwin, C. (1859). *On the origin of species.* London: Murray.

Dawkins, R. (1986). *The blind watchmaker: Why the evidence of evolution reveals a universe without design.* New York: WW Norton and Company.

de Garis, H. (1990a). Genetic programming: Building nanobrains with genetically programmed neural network modules. *In International Joint Conference on Neural Networks, 1990 IJCNN* (pp. 511–516).

de Garis, H. (1990b). Genetic programming: Building artificial nervous systems using genetically programmed neural network modules. *Proceedings of the 7th International Conference on Machine Learning, 1990* (pp. 132–139).

de Garis, H. (1990c). Genetic programming: Evolution of time dependent neural network modules which teach a pair of stick legs to walk. In *Proceedings of the 9th European Conference on Artificial Intelligence* (pp. 204–206). Stockholm, Sweden.

de Garis, H. (2005). *The Artilect War: Cosmists Vs. Terrans: A bitter controversy concerning whether humanity should build godlike massively intelligent machines.* Palm Springs, CA: ETC Publications. ISBN 0-88280-154-6.

De Jong, K. A. (2006). *Evolutionary computation: A unified approach.* Cambridge: MIT Press.

Deb, K., Pratap, A., Agarwal, S., & Meyarivan, T. (2002). A fast and elitist multiobjective genetic algorithm: NSGA-II. *IEEE Transactions on Evolutionary Computation, 6*(2), 182–197.

del Pobil, Á. P (2006)., Why do we need benchmarks in robotics research. In *International Conference on Intelligent Robot and Systems, Beijing, China*.

Diftler, M. A., Ahlstrom, T. D., Ambrose, R. O., Radford, N. A., Joyce, C. A., De La Pena, N., et al. (2012). Robonaut 2—initial activities on-board the ISS. In *IEEE Aerospace Conference, 2012* (pp. 1–12).

Dip, G., Prahlad, V., & Kien, P. D. (2009). Genetic algorithm-based optimal bipedal walking gait synthesis considering tradeoff between stability margin and speed. *Robotica, 27*(03), 355–365.

Domingues, E., Lau, N., Pimentel, B., Shafii, N., Reis, L. P., & Neves, A. J. (2011). Humanoid behaviors: From simulation to a real robot. In *Progress in Artificial Intelligence* (pp. 352–364). Berlin, Heidelberg: Springer.

Doncieux, S., Bredeche, N., & Mouret, J. B. (Eds.). (2011). *New horizons in evolutionary robotics: Extended contributions from the 2009 EvoDeRob workshop* (Vol. 341). Berlin: Springer.

Doncieux, S., Mouret, J. B., Bredeche, N., & Padois, V. (2011). Evolutionary robotics: Exploring new horizons. In *New horizons in evolutionary robotics* (pp. 3–25). Berlin, Heidelberg: Springer.

Duarte, M., Oliveira, S., & Christensen, A. L. (2012). Automatic synthesis of controllers for real robots based on preprogrammed behaviors. In *From animals to animats 12* (pp. 249–258). Berlin, Heidelberg: Springer.

Dubbin, G. A., & Stanley K. O. (2010). Learning to dance through interactive evolution. In *Proceedings of the 2010 International Conference on Applications of Evolutionary Computation—Volume Part II, EvoCOMNET'10* (pp. 331–340). Berlin, Heidelberg: Springer.

Durán, B., & Thill, S. (2012). Rob's robot: current and future challenges for humanoid robots. In Z. Riadh (Ed.), *The future of humanoid robots—research and applications*. ISBN: 978-953-307-951-6, InTech.

Eaton, M. (1993a). Process control using genetically trained neural networks. *Journal of Microcomputer Applications, 16*(2), 137–145.

Eaton, M. (1993b). Genetic algorithms and neural networks for control applications. Ph.D thesis, University of Limerick, 1993.

Eaton, M. (2007a). Explorations in evolutionary humanoid robotics. In *Proceedings of the 12th International Symposium on Artificial Life and Robotics, ISAROB, Oita, Japan* (pp. 88–91).

Eaton, M. (2007b). Evolutionary humanoid robotics: Past, present and future. In *50 Years of Artificial Intelligence: Essays Dedicated to the 50th Anniversary of Artificial Intelligence*. LNAI (Vol. 4850, pp. 42–53) Springer.

Eaton, M. (2008a). Further explorations in evolutionary humanoid robotics. *Artificial Life and Robotics, 12*(1–2), 133–137.

Eaton, M. (2008b). Evolving humanoids: Using artificial evolution as an aid in the design of humanoid robots. In I. Hitoshi (Ed.), *Frontiers in evolutionary robotics*. ISBN: 978-3-902613-19-6, InTech, DOI: 10.5772/5451.

Eaton, M. (2013). An approach to the synthesis of humanoid robot dance using non-interactive evolutionary techniques. In *IEEE International Conference on Systems, Man, and Cybernetics (SMC), 2013* (pp. 3305–3309).

Eaton, M., & Davitt, T. J. (2006). Automatic evolution of bipedal locomotion in a simulated humanoid robot with many degrees of freedom. In *Proceedings of the 11th International Symposium on Artificial Life and Robotics, ISAROB, Oita, Japan* (pp. 448–451).

Eaton, M., & Davitt, T. J. (2007). Evolutionary control of bipedal locomotion in a high degree-of-freedom humanoid robot: first steps. *Artificial Life and Robotics, 11*(1), 112–115.

Eaton, M., Collins, J. J., & Sheehan, L. (2001). Toward a benchmarking framework for research into bio-inspired hardware-software artefacts. *Artificial Life and Robotics, 5*(1), 40–45.

Eaton, M., McMillan, M., & Tuohy, M. (2002). Pursuit-evasion using evolutionary algorithms in an immersive three-dimensional environment. In *IEEE International Conference on Systems, Man and Cybernetics SMC02* (Vol. 2, pp. 348–353).

Echeverria, G., Lassabe, N., Degroote, A., & Lemaignan, S. (2011). Modular open robots simulation engine: Morse. In *IEEE International Conference on Robotics and Automation (ICRA), 2011* (pp. 46–51).

Eiben, A. E., & Smith, J. E. (2003). *Introduction to evolutionary computing*. Berlin, Germany: Springer.

Endo, K., Maeno, T., & Kitano, H. (2002). Co-evolution of morphology and walking pattern of biped humanoid robot using evolutionary computation. Consideration of characteristic of the servomotors. In *IEEE/RSJ International Conference on Intelligent Robots and Systems, 2002* (Vol. 3, pp. 2678–2683)

Endo, K., Maeno, T., & Kitano, H. (2003a). Co-evolution of morphology and walking pattern of biped humanoid robot using evolutionary computation-evolutionary designing method and its evaluation. In *Proceedings of the 2003 IEEE/RSJ International Conference on Intelligent Robots and Systems (IROS 2003)* (Vol. 1, pp. 340–345).

Endo, K., Yamasaki, F., Maeno, T., & Kitano, H. (2003b). Co-evolution of morphology and controller for biped humanoid robot. In *RoboCup 2002: Robot soccer world cup VI* (pp. 327–341). Berlin, Heidelberg: Springer.

Farchy, A., Barrett, S., MacAlpine, P., & Stone, P. (2013). Humanoid robots learning to walk faster: From the real world to simulation and back. In *Proceedings of the 2013 International Conference on Autonomous Agents and Multi-agent Systems, International Foundation for Autonomous Agents and Multiagent Systems* (pp. 39–46).

Floreano, D., & Mondada, F. (1996). Evolution of homing navigation in a real mobile robot. *IEEE Transactions on Systems, Man, and Cybernetics, Part B: Cybernetics, 26*(3), 396–407.

Floreano, D., & Urzelai, J. (2001). Evolution of plastic control networks. *Autonomous Robots, 11*(3), 311–317.

Fogel, L. J., Owens, A. J., & Walsh, M. J. (1966). *Artificial intelligence through simulated evolution*. New York: Wiley.

Friedmann, M., Petersen, K., & von Stryk, O. (2008). Simulation of multi-robot teams with flexible level of detail. In *Simulation, modeling, and programming for autonomous robots* (pp. 29–40). Berlin, Heidelberg: Springer.

Fu, Y., Moballegh, H., Rojas, R., & Jin, L. (2011). Reality Sim: A realistic environment for robot simulation platform of humanoid robot. In *2011 5th International Conference on Automation, Robotics and Applications (ICARA)* (pp. 283–287).

Fujii, A., Ishiguro, A., Aoki, T., & Eggenberger, P. (2001). Evolving bipedal locomotion with a dynamically-rearranging neural network. In *Advances in artificial life* (pp. 509–518). Berlin, Heidelberg: Springer.

Fukunaga, A., Hiruma, H., Komiya, K., & Iba, H. (2012). Evolving controllers for high-level applications on a service robot: A case study with exhibition visitor flow control. *Genetic Programming and Evolvable Machines, 13*(2), 239–263.

Fukuyama, F. (2002). *Our posthuman future: Consequences of the biotechnology revolution*. New York: Picador.

Gökçe, B., & Akin, H. L. (2011). Parameter optimization of a signal-based omni-directional biped locomotion using evolutionary strategies. In *RoboCup 2010: Robot soccer world cup XIV* (pp. 362–373). Berlin, Heidelberg: Springer.

Good, I. J. (1965). Speculations concerning the first ultraintelligent machine. *Advances in Computers, 6*(31), 88.

Gong, D., Yan, J., & Zuo, G. (2010). A review of gait optimization based on evolutionary computation. *Applied Computational Intelligence and Soft Computing, 2010*, 1–13 (Article ID 413179).

Gouaillier, D., Hugel, V., Blazevic, P., Kilner, C., Monceaux, J., Lafourcade, P., et al. (2009). Mechatronic design of NAO humanoid. In *IEEE International Conference on Robotics and Automation, 2009. ICR'09* (pp. 769–774).

Graae, C. T., Nordin, P., & Nordahl, M. (2000). Stereoscopic vision for a humanoid robot using genetic programming. In *Real-world applications of evolutionary computing* (pp. 12–21). Berlin, Heidelberg: Springer.

Grefenstette, J. J. (1986). Optimization of control parameters for genetic algorithms. *IEEE Transactions on Systems, Man and Cybernetics, 16*(1), 122–128.

Gritz, L., & Hahn, J. K. (1995). Genetic programming for articulated figure motion. *Journal of Visualization and Computer Animation, 6*(3), 129–142.

Gritz, L., & Hahn, J. K. (1997). Genetic Programming evolution of controllers for 3-D character animation. *Proceedings of Genetic Programming 1997* (pp. 139–146).

Ha, S., Han, Y., & Hahn, H. (2007). Adaptive gait pattern generation of biped robot based on human's gait pattern analysis. *International Journal of Mechanical Systems Science and Engineering, 1*(2), 80–85.

Hägele, M. et al. (2007). RoSta, Robot Standards and Reference Architectures, Deliverable D 4.1, Report on State of the Art on Benchmarks for Mobile Manipulation and Service Robots.

Hansen, N. (2006). The CMA evolution strategy: A comparing review. In *Towards a new evolutionary computation* (pp. 75–102). Berlin, Heidelberg: Springer.

Hansen, N., & Ostermeier, A. (2001). Completely derandomized self-adaptation in evolution strategies. *Evolutionary Computation, 9*(2), 159–195.

Harvey, I., Di Paolo, E. A., Wood, R., Quinn, M., & Tuci, E. (2005). Evolutionary robotics: A new scientific tool for studying cognition. *Artificial Life,11*(1–2), 79–98.

Hase, K., & Yamazaki, N. (1999). Computational evolution of human bipedal walking by a neuro-musculoskeletal mode. *Artif Life Robotics, 3*, 133–138.

Hase, K., Miyashita, K., Ok, S., & Arakawa, Y. (2003). Human gait simulation with a neuromusculoskeletal model and evolutionary computation. *Journal of Visualization and Computer Animation, 14*(2), 73–92.

Hasegawa, Y., Arakawa, T., & Fukuda, T. (2000). Trajectory generation for biped locomotion robot. *Mechatronics, 10*(1), 67–89.

Hebbel, M., Kosse, R., & Nistico, W. (2006). Modeling and learning walking gaits of biped robots. In *Proceedings of the Workshop on Humanoid Soccer Robots of the IEEE-RAS International Conference on Humanoid Robots* (pp. 40–48).

Hein, D., Hild, M., & Berger, R. (2007). Evolution of biped walking using neural oscillators and physical simulation. In *RoboCup 2007: Proceedings of the International Symposium*. Lecture Notes in Artificial Intelligence (Vol. 5001, pp. 433–440). Berlin: Springer.

Heo, J., Lee, I., & Oh, J. (2012). Development of humanoid robots in HUBO laboratory, KAIST. *Journal of the Robotics Society of Japan, 30*(4), 367–371.

Heralić, A., Wolff, K., & Wahde, M. (2007). Central pattern generators for gait generation in bipedal robots. In A. C. de Pina Filho (Ed.), *Humanoid Robots—New Developments* (pp. 285–304). Vienna, Austria: I-Tech Education and Publishing.

Hettiarachchi, D. S., & Iba, H. (2010). Evolution of a yoga performing humanoid. In *2010 Second World Congress on Nature and Biologically Inspired Computing (NaBIC)* (pp. 78–83).

Hettiarachchi, D. S., & Iba, H. (2012). An evolutionary computational approach to humanoid motion planning. *International Journal of Advanced Robotic Systems, 9*, 167.

Hoffmeister, F., & Schwefel H. P. (1990). A taxonomy of parallel evolutionary algorithms In G. Wolf, T. Legendi, & U. Schendel (Eds.), *Parcella '90, Proceedings 5th International Workshop on Parallel Processing by Cellular Automata and Arrays* (Vol. 2, pp. 97–107). Berlin: Academic Press.

Hoffmeister, F., & Bäck, T. (1991). *Genetic algorithms and evolution strategies: Similarities and differences* (pp. 455–469). Berlin, Heidelberg: Springer.

Holland, J. H. (1975). *Adaptation in natural and artificial systems*. Ann Arbor, MI: University of Michigan Press.

Hong, Y. D., Kim, Y. H., & Kim, J. H. (2009). Evolutionary optimized footstep planning for humanoid robot. In *IEEE International Symposium on Computational Intelligence in Robotics and Automation (CIRA), 2009* (pp. 266–271).

Igel, C., Hansen, N., & Roth, S. (2007). Covariance matrix adaptation for multi-objective optimization. *Evolutionary Computation, 15*(1), 1–28.

Iocchi, L., Libera, F. D., & Menegatti, E. (2007) Learning humanoid soccer actions interleaving simulated and real data. In *Proceedings of the Second Workshop on Humanoid Soccer Robots, IEEE-RAS 7th International Conference on Humanoid Robots, Pittsburgh, 2007.*

Ishiguro, A., Kawasumi, K., & Fujii, A. (2002). Increasing evolvability of a locomotion controller using a passive-dynamic-walking embodiment. In *IEEE/RSJ International Conference on Intelligent Robots and Systems, 2002* (Vol. 3, pp. 2581–2586).

Ishiguro, A., Fujii, A., & Hotz, P. E. (2003). Neuromodulated control of bipedal locomotion using a polymorphic CPG circuit. *Adaptive Behavior, 11*(1), 7–17.

Ishihara, H., Yoshikawa, Y., & Asada, M. (2011). Realistic child robot "affetto" for understanding the caregiver-child attachment relationship that guides the child development. In *2011 IEEE International Conference on Development and Learning (ICDL)* (Vol. 2, pp. 1–5).

Jackson, J. (2007). Microsoft robotics studio: A technical introduction. *IEEE Robotics and Automation Magazine, 14*(4), 82–87.

Jadhav, S., Joshi, M., & Pawar, J. (2012). Art to SMart: An evolutionary computational model for BharataNatyam choreography. In *2012 12th International Conference on Hybrid Intelligent Systems (HIS)* (pp. 384–389).

Jakobi, N. (1997a). Evolutionary robotics and the radical envelope-of-noise hypothesis. *Adaptive Behavior, 6*(2), 325–368.

Jakobi, N. (1997b). Half-baked, ad-hoc and noisy: Minimal simulations for evolutionary robotics. In: P. Husbands, & I. Harvey (Eds.), *Proceedings of the Fourth European Conference on Artificial Life*. Cambridge, MA: MIT Press.

Jeon, K. S., Kwon, O., & Park, J. H. (2004). Optimal trajectory generation for a biped robot walking a staircase based on genetic algorithms. In *Proceedings of the 2004 IEEE/RSJ International Conference on Intelligent Robots and Systems, 2004 (IROS 2004)* (Vol. 3, pp. 2837–2842)

Jingdong, Y., Bingrong, H., Songhao, P., & Qingcheng, H. (2007). An efficient strategy of penalty kick and goal keep based on evolutionary walking gait for biped soccer robot. *Information Technology Journal, 6*, 1120–1129.

Joy, B. (2000). Why the future doesn't need us. *Wired, 8*(4), 1–11.

Juárez-Guerrero, J., Munoz-Gutiérrez, S., & Cuevas, W. M. (1998). Design of a walking machine structure using evolutionary strategies. In *Proceedings of the 1998 IEEE International Conference on Systems, Man, and Cybernetics* (Vol. 2, pp. 1427–1432).

Kambayashi, Y., Takimoto, M., & Kodama, Y. (2005). Controlling biped walking robots using genetic algorithms in mobile agent environment. In *IEEE 3rd International Conference on Computational Cybernetics, 2005. (ICCC 2005)* (pp. 29–34).

Kamio, S., & Iba, H. (2004). Evolutionary construction of a simulator for real robots. In *IEEE Congress on Evolutionary Computation, 2004 (CEC 2004)* (Vol. 2, pp. 2202–2209).

Kanehiro, F., Hirukawa, H., & Kajita, S. (2004). Open architecture humanoid robotics platform. *Journal of Robotics Research, 23*(2), 155–165.

Kaneko, K., Kanehiro, F., Morisawa, M., Miura, K., Nakaoka, S., & Kajita, S. (2009). Cybernetic human HRP-4C. In *9th IEEE-RAS International Conference on Humanoid Robots, 2009. Humanoids 2009* (pp. 7–14).

Kaneko, K., Kanehiro, F., Morisawa, M., Tsuji, T., Miura, K., Nakaoka, S., et al. (2011). Hardware improvement of cybernetic human HRP-4C for entertainment use. In *Proceedings of the 2011 IEEE/RSJ International Conference on Intelligent Robots and Systems (IROS)* (pp. 4392–4399)

Karlsson, R., Nordin, P., & Nordahl, M. (2000). Sound localization for a humanoid robot by means of genetic programming. In *Real-world applications of evolutionary computing* (pp. 65–76). Berlin, Heidelberg: Springer.

Kee, D., Wyeth, G., & Roberts, J. (2004). Biologically inspired joint control for a humanoid robot. In *4th IEEE/RAS International Conference on Humanoid Robots* (Vol. 1, pp. 385–401).

Kim, D. W., de Silva, C. W., & Park, G. T. (2010). Evolutionary design of Sugeno-type fuzzy systems for modelling humanoid robots. *International Journal of Systems Science, 41*(7), 875–888.

Kim, E., Kim, M., & Kim, J. W. (2009). Optimal trajectory generation for walking up and down a staircase with a biped robot using genetic algorithm (GA). In *Advances in robotics* (pp. 103–111). Berlin, Heidelberg: Springer.

Kitano, H., & Asada, M. (1998). RoboCup humanoid challenge: That's one small step for a robot, one giant leap for mankind. In *Proceedings of the 1998 IEEE/RSJ International Conference on Intelligent Robots and Systems* (Vol. 1, pp. 419–424).

Kitano, H., Asada, M., Kuniyoshi, Y., Noda, I., Osawai, E., & Matsubara, H. (1998). Robocup: A challenge problem for AI and robotics. In *RoboCup-97: Robot soccer world cup I* (pp. 1–19). Berlin, Heidelberg: Springer.

Koenig, N., & Howard, A. (2004). Design and use paradigms for Gazebo, an open-source multi-robot simulator. In *Proceedings of the 2004 IEEE/RSJ International Conference on Intelligent Robots and Systems, 2004 (IROS 2004).* (Vol. 3, pp. 2149–2154)

Koos, S., Mouret, J. B., & Doncieux, S. (2010). Crossing the reality gap in evolutionary robotics by promoting transferable controllers. In *Proceedings of the 12th Annual Conference on Genetic and Evolutionary Computation* (pp. 119–126). ACM.

Koos, S., Mouret, J. B., & Doncieux, S. (2013). The transferability approach: Crossing the reality gap in evolutionary robotics. *IEEE Transactions on Evolutionary Computation, 17*(1), 122–145.

Koza, J. R. (1992). *Genetic programming: On the programming of computers by means of natural selection.* Cambridge: MIT Press.

Krasnow, D., & Chatfield, S. J. (2009). Development of the "Performance competence evaluation measure": Assessing qualitative aspects of dance performance. *Journal of Dance Medicine and Science, 13*(4), 101–107.

Kulk, J., & Welsh, J. S. (2011). Evaluation of walk optimisation techniques for the Nao robot. In *Proceedings of the 2011 11th IEEE-RAS International Conference on Humanoid Robots (Humanoids)* (pp. 306–311).

Kulvanit, P., Chaiyaratana, N., & Laowattana, D. (2007). Biped fast walking gait shaping via evolutionary multi-objective optimization. In *IEEE Congress on Evolutionary Computation, 2007 (CEC 2007)* (pp. 4019–4026).

Lächele, J., Franchi, A., Bülthoff, H. H., & Giordano, P. R. (2012). SwarmSimX: Real-time simulation environment for multi-robot systems. In *Simulation, modeling, and programming for autonomous robots* (pp. 375–387). Berlin, Heidelberg: Springer.

Langdon, W. B., & Nordin, P. (2001). Evolving hand-eye coordination for a humanoid robot with machine code genetic programming. In *Genetic programming* (pp. 313–324). Berlin, Heidelberg: Springer.

Laue, T., & Hebbel, M. (2009). Automatic parameter optimization for a dynamic robot simulation. In *RoboCup 2008: Robot soccer world cup XII* (pp. 121–132). Berlin, Heidelberg, Springer

Lee, J. Y., Kim, M. S., & Lee, J. J. (2004). Multi-objective walking trajectories generation for a biped robot. In *Proceedings of the 2004 IEEE/RSJ International Conference on Intelligent Robots and Systems, 2004 (IROS 2004)* (Vol. 4, pp. 3853–3858)

Lee, W. P., Jong, J. S., & Yang, T. H. (2010). Evolving behavior sequences for a humanoid entertainment robot. *Artificial Life and Robotics, 15*(3), 341–346.

Lehman, J., & Stanley, K. O. (2008). Exploiting open-endedness to solve problems through the search for novelty. In *Proceedings of the Eleventh International Conference on Artificial Life (ALIFE XI)* (pp. 329–336).

Lehman, J., & Stanley, K. O. (2011). Abandoning objectives: Evolution through the search for novelty alone. *Evolutionary Computation, 19*(2), 189–223.

Leibold, S., Fregin, A., Kaczor, D., Kollmitz, M., El Menuawy, K., Popp, E., et al. (2013). RoboCup@ work league winners 2012. In *RoboCup 2012: Robot soccer world cup XVI* (pp. 65–76). Berlin, Heidelberg: Springer.

Lewis, M. A., Fagg, A. H., & Solidum, A. (1992). Genetic programming approach to the construction of a neural network for control of a walking robot. In *Proceedings of the 1992 IEEE International Conference on Robotics and Automation* (pp. 2618–2623).

Lipson, H., Bongard, J. C., Zykov, V., & Malone, E. (2006). Evolutionary robotics for legged machines: From simulation to physical reality. In T. Arai, et al. (Eds.), *Intelligent Autonomous Systems 9 (IAS-9)* (pp. 11–18).

Liu, H., & Iba, H. (2004a). A hierarchical approach for adaptive humanoid robot control. In *IEEE Congress on Evolutionary Computation, 2004 (CEC 2004)* (Vol. 2, pp. 1546–1553).

Liu, H., & Iba, H. (2004b). A layered control architecture for humanoid robot. In *Proceedings of International Conferences on Autonomous Robots and Agents* (pp. 424–439). New Zealand.

Lungarella, M., Iida, F., Bongard, J., & Pfeifer, R. (Eds.). (2007). *50 Years of Artificial Intelligence: Essays Dedicated to the 50th Anniversary of Artificial Intelligence*. LNAI (Vol. 4850). Springer.

Massera, G., Cangelosi, A., & Nolfi, S. (2007). Evolution of prehension ability in an anthropomorphic neurorobotic arm. *Frontiers in Neurorobotics, 1*(4), 1–9.

Matarić, M., & Cliff, D. (1996). Challenges in evolving controllers for physical robots. *Robotics and Autonomous Systems, 19*(1), 67–83.

McCarthy, J., Minsky, M., Rochester, N., & Shannon, C. (1955, August 31). *A Proposal for the Dartmouth Summer Research Project on Artificial Intelligence*. Formal Reasoning Group, Stanford University, Stanford, CA.

McHale, G., & Husbands, P. (2004). Gasnets and other evolvable neural networks applied to bipedal locomotion. In *From Animals to Animats 8, Proceedings of the Eighth International Conference on the Simulation of Adaptive Behavior* (pp. 163–172).

Menzel, P., & d'Aluisio, F. (2000). *Robo sapiens: Evolution of a new species*. Boston: MIT Press.

Michel, O. (2004). Webots: Professional mobile robot simulation. *International Journal of Advanced Robotic Systems, 1*(1), 39–42.

Michel, O., Rohrer, F., & Bourquin, Y. (2008). Rat's life: A cognitive robotics benchmark. In *European Robotics Symposium* 2008 (pp. 223–232). Berlin, Heidelberg: Springer.

Miglino, O., Lund, H. H., & Nolfi, S. (1995). Evolving mobile robots in simulated and real environments. *Artificial Life, 2*(4), 417–434.

Minsky, M. (1970, November 20). Life Magazine (p. 68).

Miyashita, K., Ok, S., & Hase, K. (2003). Evolutionary generation of human-like bipedal locomotion. *Mechatronics, 13*(8), 791–807.

Mondada, F., Bonani, M., Raemy, X., Pugh, J., Cianci, C., Klaptocz, A., et al. (2009). The e-puck, a robot designed for education in engineering. In *Proceedings of the 9th Conference on Autonomous Robot Systems and Competitions* (Vol. 1(1), pp. 59–65). IPCB, Instituto Politécnico de Castelo Branco.

Mori, M. (1970). Bukimi no tani (the Uncanny Valley). *Energy, 7*, 33–35.

Nagasaka, K., Konno, A., Inaba, M., and Inoue, H. (1997). Acquisition of visually guided swing motion based on genetic algorithms and neural networks in two-armed bipedal robot. In *Proceedings of the 1997 IEEE International Conference on Robotics and Automation, 1997* (Vol. 4, pp. 2944–2949).

Nelson, A., Barlow, G., & Doitsidis, L. (2009). Fitness functions in evolutionary robotics: A survey and analysis. *Robotics and Autonomous Systems, 57*(4), 345–370.

Nelson, G., Saunders, A., Neville, N., Swilling, B., Bondaryk, J., Billings, D., et al. (2012). Petman: A humanoid robot for testing chemical protective clothing. *Journal of the Robotics Society of Japan, 30*(4), 372–377.

Nof, S. Y. (Ed.) (2009). *Springer handbook of automation*. Berlin: Springer.

Nolfi, S., & Floreano, D. (2000). Evolutionary robotics. *The biology, intelligence, and technology of self-organizing machines*. Cambridge: MIT Press.

Obst, O., & Rollmann, M. (2005). Spark—A generic simulator for physical multi-agent simulations. *Computer Systems Science and Engineering, 20*(5), 347–356.

Ogihara, N., & Yamazaki, N. (2001). Generation of human bipedal locomotion by a bio-mimetic neuro-musculo-skeletal model. *Biological Cybernetics, 84*, 1–11.

Oh, J. H., Hanson, D., Kim, W. S., Han, I. Y., Kim, J. Y., & Park, I. W. (2006). Design of android type humanoid robot Albert HUBO. In *IEEE/RSJ International Conference on Intelligent Robots and Systems, 2006* (pp. 1428–1433).

Ok, S., Miyashita, K., & Hase, K. (2001). Evolving bipedal locomotion with genetic programming–a preliminary report. In *IEEE Proceedings of the 2001 Congress on Evolutionary Computation* (Vol. 2, pp. 1025–1032).

Ouannes, N., Djedi, N., Duthen, Y., & Luga, H. (2012). Gait evolution for humanoid robot in a physically simulated environment. In *Intelligent computer graphics 2011* (pp. 157–173). Berlin, Heidelberg: Springer.

Palmer, M., Miller, D., & Blackwell, T. (2009). An evolved neural controller for bipedal walking: Transitioning from simulator to hardware. In *Proceedings of IROS 2009 Workshop on Exploring New Horizons in Evolutionary Design of Robots*.

Park, J. H., & Choi, M. (2004). Generation of an optimal gait trajectory for biped robots using a genetic algorithm. *JSME International Journal Series C, 47*(2), 715–721.

Paul, C., & Bongard, J. C. (2001). The road less travelled: Morphology in the optimization of biped robot locomotion. In *Proceedings of the 2001 IEEE/RSJ International Conference on Intelligent Robots and Systems* (Vol. 1, pp. 226–232)

Pettersson, J., Sandholt, H., & Wahde, M. (2001). A flexible evolutionary method for the generation and implementation of behaviors for humanoid robots. In *Proceedings of the IEEE-RAS International Conference on Humanoid Robots* (pp. 279–286).

Pfeifer, R., Bongard, J., & Grand, S. (2007). *How the body shapes the way we think: A new view of intelligence*. Cambridge: MIT Press.

Picado, H., Gestal, M., Lau, N., Reis, L. P., & Tomé, A. M. (2009). Automatic generation of biped walk behavior using genetic algorithms. In *Bio-inspired systems: Computational and ambient intelligence* (pp. 805–812). Berlin, Heidelberg: Springer.

Pinciroli, C., Trianni, V., O'Grady, R., Pini, G., Brutschy, A., Brambilla, M., et al. (2012). ARGoS: A modular, multi-engine simulator for heterogeneous swarm robotics. In *Proceedings of the 2011 IEEE/RSJ International Conference on Intelligent Robots and Systems (IROS)* (pp. 5027–5034).

Ra, S., Park, G., Kim, C. H., & You, B. J. (2008). PCA-based genetic operator for evolving movements of humanoid robot. In *IEEE Congress on Evolutionary Computation, 2008. CEC 2008 (IEEE World Congress on Computational Intelligence)* (pp. 1219–1225).

Raibert, M., Blankespoor, K., Nelson, G., & Playter, R. (2008). Bigdog, the rough-terrain quadruped robot. In *Proceedings of the 17th International Federation of Automatic Control (IFAC) World Congress* (pp. 10823–10825).

Rechenberg, I. (1973). *Evolutionsstrategie: Optimierung technischer systeme nach Prinzipien der biologischen evolution*. Stuttgart: Frommann-Holzboog Verlag.

Reil, T., & Husbands, P. (2002). Evolution of central pattern generators for bipedal walking in a real-time physics environment. *IEEE Transactions on Evolutionary Computation, 6*(2), 159–168.

Röfer T., Mayer N., Savage J., & Saranli U. (Eds.). (2012). *RoboCup 2011: Robot Soccer World Cup XV (papers from the 15th Annual RoboCup International Symposium, Istanbul, Turkey, July 2011)*. Lecture Notes in Computer Science (Vol. 7416). Springer

Sakai, M., Kanoh, M., & Nakamura, T. (2012). Evolutionary multivalued decision diagrams for obtaining motion representation of humanoid robots. *IEEE Transactions on Systems, Man, and Cybernetics, Part C: Applications and Reviews, 42*(5), 653–663.

Sandini, G., Metta, G., & Vernon, D. (2007). The iCub cognitive humanoid robot: An open-system research platform for enactive cognition. In M. Lungarella (Ed.), *50 Years of AI* (pp. 358–369). Berlin, Germany: Springer.

Santos, J. (2013). Evolved center-crossing recurrent synaptic delay based neural networks for biped locomotion control. In *Proceedings of the 2013 IEEE Congress on Evolutionary Computation (CEC)* (pp. 142–148).

Santos, J., & Campo, Á. (2012). Biped locomotion control with evolved adaptive center-crossing continuous time recurrent neural networks. *Neurocomputing, 86,* 86–96.

Sato, T., Matsuhira, N., & Oyama, E. (2008). Common platform technology for next-generation robots. In *Workshop on Standard and Common Platform for Robotics, International Conference on Simulation, Modeling and Programming for Autonomous Robots* (pp. 616–627).

Savastano, P., & Nolfi, S. (2012). Incremental learning in a 14 DOF simulated iCub robot: Modeling infant reach/grasp development. In *Biomimetic and Biohybrid Systems* (pp. 250–261). Berlin, Heidelberg: Springer.

Schneider, F. E., & Wildermuth, D. (2011). Results of the European land robot trial and their usability for benchmarking outdoor robot systems. In *Towards autonomous robotic systems* (pp. 408–409). Berlin, Heidelberg: Springer.

Schneider, F. E., Wildermuth, D., & Wolf, H. (2012). Professional ground robotic competitions from an educational perspective: A consideration using the example of the European Land Robot Trial (ELROB). In *Proceeding of the 2012 6th IEEE International Conference Intelligent Systems (IS)* (pp. 399–405).

Schreiner, D., & Punzengruber, C. (2011). Parametrizing Motion Controllers of Humanoid Robots by Evolution, INFORMATIK 2011 - Informatik schafft Communities 41. Jahrestagung der Gesellschaft für Informatik, 4.-7.10.2011, Berlin.

Sellers, W. I., & Manning, P. L. (2007). Estimating dinosaur maximum running speeds using evolutionary robotics. *Proceedings of the Royal Society B: Biological Sciences, 274*(1626), 2711–2716.

Sellers, W. I., Dennis, L. A., Wang, W. J., & Crompton, R. H. (2004). Evaluating alternative gait strategies using evolutionary robotics. *Journal of Anatomy, 204*(5), 343–351.

Sellers, W. I., Cain, G., Wang, W. J., & Crompton, R. H. (2005). Stride lengths, speed and energy costs in walking of Australopithecus afarensis: Using evolutionary robotics to predict locomotion of early human ancestors. *Journal of the Royal Society Interface, 2,* 431–442.

Sellers, W. I., Pataky, T. C., Caravaggi, P., & Crompton, R. H. (2010). Evolutionary robotic approaches in primate gait analysis. *International Journal of Primatology, 31,* 321–338.

Shafii, N., Aslani, S., Nezami, O. M., & Shiry, S. (2010). Evolution of biped walking using truncated Fourier series and particle swarm optimization. In *RoboCup 2009: Robot soccer world cup XIII* (pp. 344–354). Berlin, Heidelberg: Springer.

Sheng, B., Huaqing, M., Qifeng, L., & Xijing, Z. (2009). Multi-objective optimization for a humanoid robot climbing stairs based on genetic algorithms. In *International Conference on Information and Automation, 2009. ICIA'09* (pp. 66–71)

Shrivastava, M., Dutta, A., & Saxena, A. (2007). Trajectory generation using GA for an 8 DOF biped robot with deformation at the sole of the foot. *Journal of Intelligent and Robotic Systems, 49*(1), 67–84.

Siciliano, B., & Khatib, O. (Eds.). (2008). *Springer handbook of robotics.* Berlin: Springer. ISBN 978-3-540-23957-4.

Sims, K. (1994a). Evolving virtual creatures. In *Proceedings of the 21st Annual Conference on Computer Graphics and Interactive Techniques* (pp. 15–22). ACM.

Sims, K. (1994b). Evolving 3D morphology and behavior by competition. In R. Brooks, & P. Maes, (Eds.), *Proceedings of artificial life IV* (pp. 28–39). Cambridge, MA: MIT Press.

Stanley, K. O. (2011). Why evolutionary robotics will matter. In *New horizons in evolutionary robotics* (pp. 37–41). Berlin, Heidelberg: Springer.

Stanley, K. O., & Miikkulainen, R. (2002). Evolving neural networks through augmenting topologies. *Evolutionary Computation, 10*(2), 99–127.

Stanley, K. O., & Miikkulainen, R. (2004). Competitive coevolution through evolutionary complexification. *Journal of Artificial Intelligence Research (JAIR), 21*, 63–100.

Stanley, K. O., Bryant, B. D., & Miikkulainen, R. (2005). Real-time neuroevolution in the NERO video game. *IEEE Transactions on Evolutionary Computation, 9*(6), 653–668.

Stückler, J., Badami, I., Droeschel, D., Gräve, K., Holz, D., McElhone, M., et al. (2013). Nimbro@ home: Winning team of the RoboCup@ home competition 2012. In *RoboCup 2012: Robot soccer world cup XVI* (pp. 94–105). Berlin, Heidelberg: Springer.

Suzuki, M., Gritti, T., & Floreano, D. (2009). Active vision for goal-oriented humanoid robot walking. In *Creating brain-like intelligence* (pp. 303–313). Berlin, Heidelberg: Springer.

Takagi, H. (2001). Interactive evolutionary computation: Fusion of the capabilities of EC optimization and human evaluation. *Proceedings of the IEEE, 89*(9), 1275–1296.

Tang, Z., Zhou, C., & Sun, Z. (2005). Humanoid walking gait optimization using GA-based neural network. In *Advances in natural computation* (pp. 252–261). Berlin Heidelberg: Springer.

Tikhanoff, V., Fitzpatrick, P., Nori, F., Natale, L., Metta, G., & Cangelosi, A. (2008, September). The iCub humanoid robot simulator. In *IROS Workshop on Robot Simulators* (Vol. 22).

Tikhanoff, V., Cangelosi, A., Fitzpatrick, P., Metta, G., Natale, L., & Nori, F. (2008). An open-source simulator for cognitive robotics research: The prototype of the iCub humanoid robot simulator. In *Proceedings of the 8th Workshop on Performance Metrics for Intelligent Systems* (pp. 57–61). ACM.

Tikhanoff, V., Cangelosi, A., & Metta, G. (2011). Integration of speech and action in humanoid robots: iCub simulation experiments. *IEEE Transactions on Autonomous Mental Development, 3*(1), 17–29.

Torres, E., & Garrido, L. (2012). Automated generation of CPG-based locomotion for robot Nao. In *RoboCup 2011: Robot soccer world cup XV* (pp. 461–471). Berlin, Heidelberg: Springer.

Torricelli, D., Mizanoor, R. S., Gonzalez, J., Lippi, V., Hettich, G., Asslaender, L., et al. (2014). Benchmarking human-like posture and locomotion of humanoid robots: A Preliminary scheme. In *Biomimetic and biohybrid systems* (pp. 320–331). Berlin: Springer International Publishing.

Tuci, E., Massera, G., & Nolfi, S. (2010). Active categorical perception of object shapes in a simulated anthropomorphic robotic arm. *IEEE Transactions on Evolutionary Computation, 14*(6), 885–899.

Turing, A. M. (1950). Computing machinery and intelligence. *Mind, 59*, 433–460.

Uchiyama, T., Morita, T., & Sawasaki, N. (2011). Development of personal robot. In *Robotics research* (pp. 319–336). Berlin, Heidelberg: Springer.

Urieli, D., MacAlpine, P., Kalyanakrishnan, S., Bentor, Y., & Stone, P. (2011). On optimizing interdependent skills: A case study in simulated 3D humanoid robot soccer. In *The 10th International Conference on Autonomous Agents and Multiagent Systems, International Foundation for Autonomous Agents and Multiagent Systems* (Vol. 2, pp. 769–776).

van Noort, S., & Visser, A. (2012). Validation of the dynamics of an humanoid robot in USARSim. In *Proceedings of the Workshop on Performance Metrics for Intelligent Systems* (pp. 190–197). ACM.

Vaughan, E., Di Paolo, E. A., & Harvey, I. (2004). The evolution of control and adaptation in a 3D powered passive dynamic walker. In *Proceedings of the Ninth International Conference on the Simulation and Synthesis of Living Systems, Artificial Life IX* (pp. 139–145).

Vaughan, E., Di Paolo, E., & Harvey, I. (2005). The tango of a load balancing biped. In *Climbing and walking robots* (pp. 813–823). Berlin, Heidelberg: Springer.

Vaughan, R. (2008). Massively multi-robot simulation in stage. *Swarm Intelligence, 2*(2–4), 189–208.

Veloso, M., & Stone, P. (2012). Video: RoboCup robot soccer history 1997–2011. In *Proceedings of the IEEE/RSJ International Conference on Intelligent Robots and Systems (IROS)* (pp. 5452–5453).

Virčíková, M., & Sinčák, P. (2009). Dance choreography design of humanoid robots using interactive evolutionary computation. In *Human-friendly robotics: 3rd Workshop for Young Researchers*. Tübingen, Germany: Max Planck Institute for Biological Cybernetics.

Virčíková, M., & Sinčák, P. (2011). Discovering art in robotic motion: From imitation to innovation via interactive evolution. In *Ubiquitous computing and multimedia applications* (pp. 183–190). Berlin, Heidelberg: Springer.

Vukobratovic, M., & Borovac, B. (2004). Zero-moment point—thirty five years of its life. *International Journal of Humanoid Robotics, 1*(01), 157–173.

Wakaki, H., Tokui, N., & Iba, H. (2002). Motion design of a 3D-CG avatar using interactive evolutionary computation. In *Proceedings of the 2002 IEEE International Conference on Systems, Man and Cybernetics* (Vol. 4, p. 6).

Wampler, K., & Popović, Z. (2009). Optimal gait and form for animal locomotion. *ACM Transactions Graph, 28*(3), 60.

Wang, J. M., Hamner, S. R., Delp, S. L., & Koltun, V. (2012). Optimizing locomotion controllers using biologically-based actuators and objectives. *ACM Transactions Graph, 31*(4), 25.

Wang, X., Lu, T., & Zhang, P. (2008). State generation method for humanoid motion planning based on genetic algorithm. *Journal of Humanoids, 1*(1), 17–24.

Wang, X., Lu, T., & Zhang, P. (2012). State generation method for humanoid motion planning based on genetic algorithm. *International Journal of Advanced Robotic Systems, 9*, 1–9.

Wiener, N. (1948). *Cybernetics: or control and communication in the animal and the machine.* New York: Wiley.

Wolff, K., & Nordin, P. (2001). Evolution of efficient gait with humanoids using visual feedback. In *Proceedings of the 2nd IEEE-RAS International Conference on Humanoid Robots, Humanoids* (pp. 99–106).

Wolff, K., & Nordin, P. (2002). Evolution of efficient gait with an autonomous biped robot using visual feedback. In *Proceedings of the 8th Mechatronics Forum International Conference. University of Twente, Enschede, the Netherlands* (pp. 504–513).

Wolff, K., & Nordin, P. (2003a). An evolutionary based approach for control programming of humanoids. In *Proceedings of the 3rd International Conference on Humanoid Robots (Humanoids' 03) (Karlsruhe, Germany), IEEE, VDI/VDE-GMA*.

Wolff, K., & Nordin, P. (2003b). Learning biped locomotion from first principles on a simulated humanoid robot using linear genetic programming. In *Genetic and Evolutionary Computation— GECCO 2003* (pp. 495–506). Berlin, Heidelberg: Springer.

Wolff, K., Sandberg, D., & Wahde, M. (2008). Evolutionary optimization of a bipedal gait in a physical robot. In *IEEE Congress on Evolutionary Computation, 2008. CEC 2008. (IEEE World Congress on Computational Intelligence)* (pp. 440–445).

Wu, J. C., & Popović, Z. (2010 July). Terrain-adaptive bipedal locomotion control. *ACM Transactions on Graphics, 29*(4), 72:1–72:10.

Yanase, T., & Iba, H. (2006). Evolutionary motion design for humanoid robots. In *Proceedings of the 8th Annual Conference on Genetic and Evolutionary Computation* (pp. 1825–1832). ACM.

Yanase, T., & Iba, H. (2008a). Evolutionary motion design for humanoid robots. In I. Hitoshi (Ed.), *Frontiers in evolutionary robotics*. ISBN: 978-3-902613-19-6, InTech, DOI: 10.5772/5473.

Yanase, T., & Iba, H. (2008b). Evolutionary multi-objective optimization for biped walking. In *Simulated evolution and learning* (pp. 635–644). Berlin, Heidelberg: Springer.

Yang, L., Chew, C. M., Zielinska, T., & Poo, A. N. (2007). A uniform biped gait generator with offline optimization and online adjustable parameters. *Robotica, 25*(5), 549–565.

Zagal, J. C., & Ruiz-Del-Solar, J. (2007). Combining simulation and reality in evolutionary robotics. *Journal of Intelligent and Robotic Systems, 50*(1), 19–39.

Zagal, J. C., Ruiz-del-Solar, J., & Vallejos, P. (2004). Back to reality: Crossing the reality gap in evolutionary robotics. In *IAV 2004 the 5th IFAC Symposium on Intelligent Autonomous Vehicles, Lisbon, Portugal*.

Zamiri, A., Farzad, A., Saboori, E., Rouhani, M., Naghibzadeh, M., & Fard A. (2008). An evolutionary gait generator with online parameter adjustment for humanoid robots. In *IEEE/ ACS International Conference on Computer Systems and Applications, 2008. AICCSA 2008* (pp. 9–14)

Zhang, R., & Vadakkepat, P.(2003). An evolutionary algorithm for trajectory based gait generation of biped robot. In *Proceedings of the International Conference on Computational Intelligence, Robotics and Autonomous Systems*.

The manufacturer's authorised representative in the EU is Springer
Nature Customer Service Centre GmbH, Europaplatz 3, 69115 Heidelberg,
Germany. If you have any concerns regarding our products, please
contact ProductSafety@springernature.com

Printed and bound by CPI Group (UK) Ltd, Croydon, CR0 4YY

23/04/2026

02095594-0016